Building Code Basics: Commercial

Based on the 2012 International Building Code®

Building Code Basics: Commercial

Based on the 2012 International Building Code®

International Code Council
Steve Thomas, CBO

DELMAR
CENGAGE Learning

Australia • Brazil • Japan • Korea • Mexico • Singapore • Spain • United Kingdom • United States

DELMAR
CENGAGE Learning

Building Code Basics: Commercial Based on the 2012 International Building Code®

Steve Thomas

Vice President, Technology and Trades Professional Business Unit: Gregory L. Clayton

Director of Building Trades: Taryn Zlatin McKenzue

Executive Editor: Robert Person

Associate Product Manager: Nobina Preston

Director of Marketing: Beth A. Lutz

Senior Marketing Manager: Marissa Maiella

Marketing Coordinator: Rachael Torres

Senior Production Director: Wendy Troeger

Production Manager: Sherondra Thedford

Senior Content Project Manager: Stacey Lamodi

Senior Art Director: Benjamin Gleeksman

ICC Staff:

Executive Vice President and Director of Business Development: Mark A. Johnson

Senior Vice President, Business and Product Development: Hamid Naderi

Vice President and Technical Director, Product Development: Doug Thornburg

Manager, Project and Special Sales: Suzane Nunes

Senior Marketing Specialist: Dianna Hallmark

Library of Congress Control Number: 2012936300

ISBN-13: 978-1-133-28335-5

ICC World Headquarters
500 New Jersey Avenue, NW
6th Floor
Washington, D.C. 20001-2070
Telephone: 1-888-ICC-SAFE (422-7233)
Website: http://www.iccsafe.org

Delmar
5 Maxwell Drive
Clifton Park, NY 12065-2919
USA

Cengage Learning is a leading provider of customized learning solutions with office locations around the globe, including Singapore, the United Kingdom, Australia, Mexico, Brazil and Japan. Locate your local office at: **international.cengage.com/region**

Cengage Learning products are represented in Canada by Nelson Education, Ltd.

For more learning solutions, please visit our corporate website at **www.cengage.com**

Visit us at **www.InformationDestination.com**

Printed in the United States of America
1 2 3 4 5 14 13 12

CONTENTS

PART II: BUILDING PLANNING

PART III: FIRE SAFETY

PART V: HEALTH SAFETY

PART VI: STRUCTURAL SAFETY

PREFACE

Construction of buildings has progressed over the centuries from stone buildings to the steel and concrete monuments to the industrial revolution and technology today. People have an expectation that when they enter a building, it has been constructed in such a manner that if an emergency situation occurs in the building, they will be protected. As new building practices improve and new building technologies are developed, the complexity of the buildings has increased. As these complexities increase, building regulations must keep up with the technology. The codes adopted by governmental agencies should not prohibit or limit the use of new materials or technologies. However, buildings must be constructed to be safe for the occupants. Therefore, the building code has had to keep up with the complex building practices and has become a complex document in itself. The codes and standards used to regulate the construction of buildings can overwhelm the casual user of the code. Navigating the complex and detailed provisions of the code can be daunting. Not all of the provisions apply to every building.

This text, *Building Code Basics: Commercial* has been written to provide an easy-to-read overview of the *International Building Code®* (IBC). It is an illustrated look at the most common code provisions. This text provides an overall understanding of the provisions of the building code. It is presented in a user-friendly manner with emphasis on the technical accuracy and clear noncode language. This book targets readers who have a basic understanding of architecture and construction but a limited knowledge of the code requirements.

Anyone who is involved in the design, construction, or regulatory aspects of building construction can gain from this book. People with all levels of experience can profit from the information provided. The user will be able to use this document for learning the fundamental provisions and intents of the building code. The most common and most used provisions of the *2012 International Building Code®* have been addressed.

The information in *Building Code Basics: Commercial* has been organized by the different purposes of the code. It discusses the administrative provision that building departments use to enforce the provisions of the building code. The provisions used to design the overall building use, construction, size, and specific occupancy requirements are covered in the building planning portion of the code. The fire safety topics include the use of passive and active fire protection. The discussion of life safety issues includes the methods used in the design and construction to ensure that occupants are provided with a safe means of evacuating a building in an emergency. It also provides the provisions that allow people with physical disabilities to access buildings. Health safety provisions regarding weather protection and interior environment have been provided. Finally, the structural provisions of the code have been put into easy-to-understand language.

This book is not intended to cover all of the provisions of the building code or all of the acceptable materials and methods of construction. This is not to say that the information *not* included in this book is not important. This book should be used with the *2012 International Building Code*, which should be referenced for more details and information. This book only provides a basic understanding of a very complex document and referenced standards.

Building Code Basics: Commercial includes full-color photos and illustrations to help the reader understand the provisions and application of the building code requirements. It provides examples, simplified tables, and highlights to explain the fundamental requirements of the *International Building Code* and to gain compliance with its regulations. References to the applicable sections or standards have been provided to assist the reader in locating more detailed and complete information in the code. A glossary of code and construction terms clarifies the meaning of the technical provisions.

ABOUT THE *INTERNATIONAL BUILDING CODE*

The IBC covers all buildings except detached one- and two-family dwellings and townhouses not more than three stories in height. This comprehensive code features time-tested safety concepts; structural, fire, and life safety provisions covering means of egress; interior finish requirements; comprehensive roof provisions; seismic engineering provisions; innovative construction technology; occupancy classifications; and the latest industry standards in material design. It is founded on broad-based principles that make possible the use of new materials and new building designs.

The IBC is one of the codes in the family of *International Codes* published by the International Code Council® (ICC). All of these codes are maintained and updated through an open code-development process and are available internationally for adoption by the governing authority to provide consistent enforceable regulation for the built environment.

ACKNOWLEDGMENTS

Building Code Basics: Commercial is the result of many hours of research and collaborative effort. The author would like to thank those people who provided valuable assistance and contribution to developing this book. The knowledge and experience of the ICC staff were instrumental in its development. Hamid Naderi, PE, Vice President of Product Development, came up with the concept of the book and provided the initial direction. Doug Thornburg, AIA, Technical Director of Product Development, provided his services as the technical editor during the writing. Doug was also a friend and provided moral support during the long nights and weekends. Finally, thanks go to Scott Stookey and Kermit Robinson, senior technical staff, for their assistance with the fire protection, special occupancies, and building planning provisions. All of these people contributed to the accuracy and quality of this document.

The author would like to dedicate this book to Perry (Pete) Tyree, PE. Pete mentored me as a young building official and taught me how to interpret and use the codes in a reasonable and fair way. He served as president of the International Conference of Building Officials (ICBO) in 1972 while serving as the building official for the Pikes Peak Regional Building Department in Colorado Springs, Colorado, for 18 years. He has continued to support the ideals and importance of the model code concept ever since. It is because of Pete Tyree that I have been able to get where I am now and have experienced the success I have enjoyed. He provided the guidance and support that I needed to become the code official and code consultant that I have become. Thank you, Pete!

ABOUT THE AUTHOR

Steve Thomas, CBO
Colorado Code Consulting, LLC
President

Mr. Thomas has over 30 years experience in building code plan review, inspection, and administration. He is currently the president of Colorado Code Consulting, LLC. He served as the Building Official for the City of Glendale, Colorado, for 18 years. Starting a building code consulting firm in 1999, his firm provides building department services for governmental agencies, code consulting services for architectural clients and building code education across the country. Mr. Thomas served on ICBO Code Development Committees and the ICBO Chairman's Advisory Committee for Small Jurisdictions. He currently serves the Chair of the ICC Means of Egress Code Development Committee. Mr. Thomas received his bachelor of science in business administration degree from the University of Phoenix. Mr. Thomas is a Certified Building Official and holds seven other ICC certifications.

About the International Code Council

The International Code Council® (ICC®) is a member-focused association dedicated to helping the building safety community and construction industry provide safe, sustainable and affordable construction through the development of codes and standards used in the design, build and compliance process. Most U.S. communities and many global markets choose the International Codes®. ICC Evaluation Service (ICC-ES), a subsidiary of the International Code Council, has been the industry leader in performing technical evaluations for code compliance fostering safe and sustainable design and construction.

Headquarters:
500 New Jersey Avenue, NW, 6th Floor
Washington, DC 20001-2070

District Offices:
Birmingham, AL; Chicago, IL; Los Angeles, CA

1-888-422-7233
www.iccsafe.org

PART

I

Code Administration and Enforcement

© International Code Council

CHAPTER 1

Introduction to Building Codes

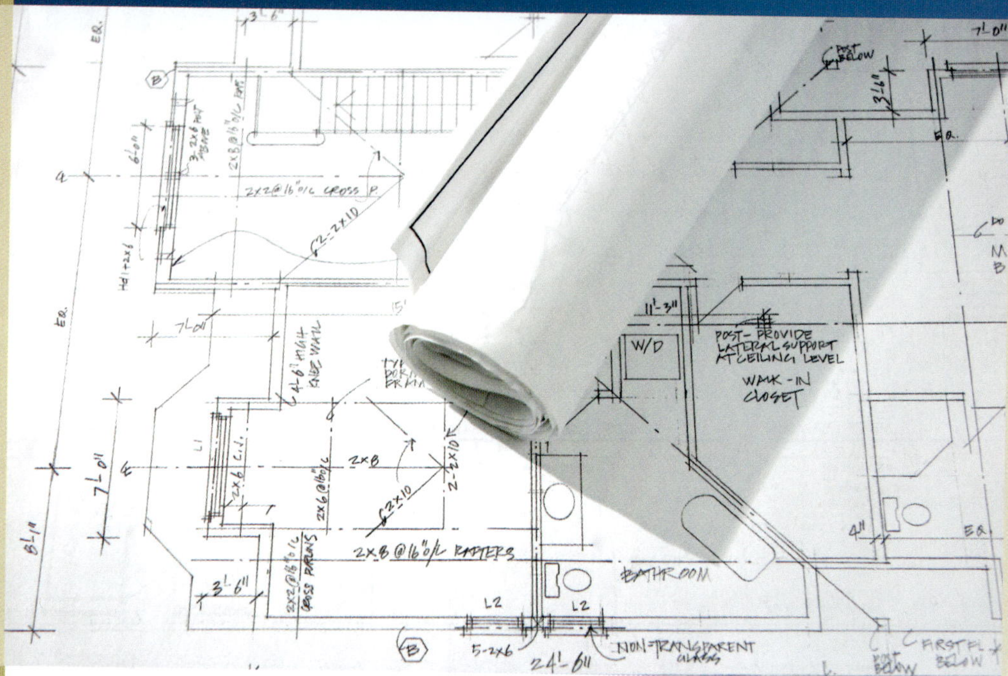

© International Code Council

Building codes are sets of regulations adopted by governmental agencies to ensure that buildings are built in a safe manner. People have an expectation that when they enter a building they will be safe from inherent dangers caused by natural or man-made disasters. We expect that if a storm passes through our area, the building will keep us dry and withstand the forces created by the storm such as wind. When the earth shakes in an earthquake, we expect the building to resist the forces and stay standing to a point that we can get out of the building. When a fire occurs in a building, the occupants need a safe and expedient way out of the building. Building codes provide these protections by limiting the potential hazards in a building and requiring certain design requirements to provide the occupants with a safe environment in which to live and work. The *International Building Code* (IBC) provides design guidelines to provide for the safety of the people using buildings across the country. The IBC regulates the architectural and structural portions of the building. It then references companion International Codes for areas such as energy, mechanical, plumbing, and electrical installations. This chapter discusses the history of codes and how they have been developed over time. It also provides insight into the administrative enforcement and application of the code on structures.

HISTORY OF CODES

Building codes have been around for ages. The first known building code was found in the *Code of Hammurabi*, created about 1760 BC in ancient Babylon (see Figure 1-1). This code, enacted by the sixth Babylonian king Hammurabi, set forth six rules for construction:

- 228. If a builder has built a house for a man, and finished it, he shall pay him a fee of two shekels of silver, for each SAR built on.
- 229. If a builder has built a house for a man, and has not made his work sound, and the house he built has fallen, and caused the death of its owner, that builder shall be put to death.
- 230. If it is the owner's son that is killed, the builder's son shall be put to death.
- 231. If it is the slave of the owner that is killed, the builder shall give slave for slave to the owner of the house.
- 232. If he has caused the loss of goods, he shall render back whatever he has destroyed. Moreover, because he did not make sound the house he built, and it fell, at his own cost he shall rebuild the house that fell.
- 233. If a builder has built a house for a man, and has not keyed his work, and the wall has fallen, that builder shall make that wall firm at his own expense.

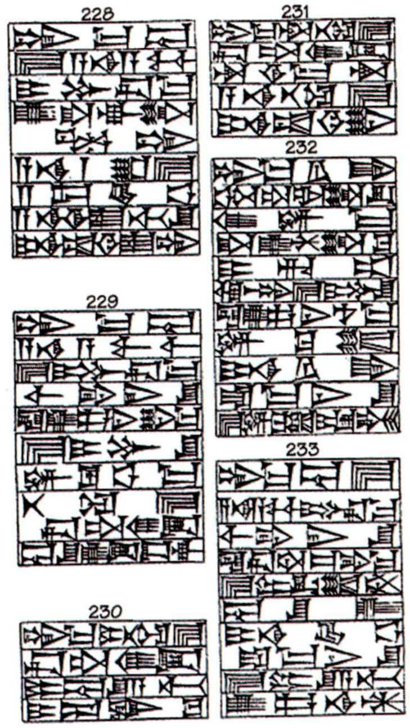

FIGURE 1-1 *Code of Hammurabi,* 1760 BC

One of the first model building codes written was the *Building Code Recommended by the National Board of Fire Underwriters* (see Figure 1-2). This code provided uniform language from a well-respected organization that could provide local governments with a model code that could be adopted without having to write one from scratch. In the 1920s, other organizations started developing model building codes. Eventually, by the end of the twentieth century, there were three model building code groups in the United States: Building Officials & Code Administrators International, Inc. (BOCA); International Conference of Building Officials, Inc. (ICBO); and Southern Building Code Congress, Inc. (SBCCI). These organizations published the *BOCA National Building Code*, *Uniform Building Code*, and *Standard Building Code*, respectively. In 1994, the codes were primarily regional in nature and their use was limited to primarily the United States. These three organizations agreed in principle to develop one model code for the country and formed the International Code Council (ICC). The first *International Building Code* developed by the ICC was published in 2000. The IBC is used in every state in the country and in some form in many other countries.

FIGURE 1-2 *1905 Building Code* Recommended by the National Board of Fire Underwriters

Code Basics

Three-Year Code Development Cycle

1. Anyone can submit a code change proposal.
2. Proposals are published and distributed for review by interested parties.
3. Open public hearings are held before code committee.
4. Public hearing results are printed and distributed.
5. Anyone can submit public comment to committee findings.
6. Public comments are published and distributed.
7. An open public final action hearing is held.
8. ICC government members vote on final action items.
9. Approved changes are incorporated into new edition of the code. ●

CODE DEVELOPMENT

As new technology develops improved building materials and products, the codes must keep up with them. We also learn from major disasters in our history. The Chicago fire in 1871 revealed that we needed to protect buildings from spread of fire from neighboring buildings. The San Francisco earthquake and subsequent fire in 1906 exposed the need for more earthquake-resistant construction. Hurricane Andrew, which hit the southern coast of Florida in 1992, brought about sweeping changes in wind-resistant construction. The attacks on the World Trade Center in 2001 also created some changes to the way high-rise buildings are built. As a result of these attacks, we learned many things about firefighter's response time and protection in high rise buildings. These types of disasters bring about changes to building codes so that the damage caused by such disasters can be reduced.

The IBC is revised and published every 3 years. Changes to the code language are submitted by anyone interested in the development of the code. The proposed changes are initially reviewed by several ICC committees comprised of experts in the field of building design, construction, and regulation (see Figure 1-3). The committees hear testimony from interested parties at a public hearing open to the public. Depending on the subject, testimony can last from a few minutes to an hour or so. Anyone can voice his or her opinion on a proposal to the committee. The committee then considers the testimony and develops a position on the proposal, either approval or disapproval. The ICC membership present at the public hearing has the opportunity to object to the findings of the committee at that time by making a motion and voting at the hearing. A code change proposal that receives a majority vote in opposition to the committee's position is automatically sent to the second step in the process of considering a code change.

The committee's decisions are then published and made available to those people interested in the code change process. Anyone can then submit a public comment objecting to the committee's position or to recommend a modification to the original proposal. The public comments and items objected to at the public hearing are published and distributed to the ICC membership and interested parties. A final action hearing is then held based on the public comments submitted. Once again, any interested parties can testify on the proposed change. After the testifying is complete, ICC governmental members vote on the merit of the proposed change. The membership can then approve, disapprove, or modify a proposal. Code change proposals that are approved by the committee and not challenged and items that are approved by the membership are then incorporated into the next edition of the code.

FIGURE 1-3 Code development committee hearing

CONSTRUCTION CODES: SCOPE AND LIMITATIONS

All of the International Codes begin with a scope and purpose of the particular code. The *scope* outlines the limitation of the code as it relates to the specific subject that it covers. It also provides a foundation for the code requirements in the specific section. The *purpose* section of the code outlines why the code has been developed. It provides the overall intent of the code as it relates to buildings. It is important to read and become familiar with the scope and purpose of the code as well as the specific sections. The general intent of construction codes is to protect the health, safety, and welfare of the people who occupy and use buildings. Each code provides this protection in different ways. In order to provide a full family of codes that can be adopted by government entities, the International Code Council publishes several different codes to address other aspects of building construction.

International Building Code

The *International Building Code* (IBC) applies to the construction, alteration, movement, enlargement, replacement, repair, equipment, use and occupancy, location, maintenance, removal, and demolition of every building or structure or any appurtenances connected or attached to such buildings or structures. The code is a set of minimum standards to protect and safeguard the public health, safety, and general welfare. It provides this protection by regulating the structural strength of buildings by outlining the loads imposed on the building and stability requirements; the means of egress facilities of buildings by providing a safe and efficient way out of a building; and a healthy environment by regulating materials and installation of those materials in a building, as well as adequate light and ventilation. This provides safety to life and property from fire and other hazards attributed to the built environment. The code also intends to provide safety to firefighters and emergency responders during emergency operations. **[Ref. 101.2, 101.3]**

Although the IBC applies to all buildings, an exception exempts detached one- and two-family dwellings and multiple single-family dwellings, also known as *townhouses*. This exception is limited to these types of structures that are no more than three stories in height with a separate means of egress and their accessory structures. Buildings that fall under this exception must comply with the *International Residential Code*. **[Ref. 101.2]**

The IBC references other codes published by the ICC. These codes are considered a part of the requirements in the IBC, but only to the extent that they are specifically referenced. The reference to the other codes does not automatically adopt those codes in their entirety. These codes address the installation of electrical, gas, mechanical, plumbing, property maintenance, fire prevention, and energy. The codes that follow provide the regulations for these areas. **[Ref. 101.4]**

International Mechanical Code

The *International Mechanical Code* (IMC) has been developed to apply to the design, installation, maintenance, alteration, and inspection of permanent mechanical systems that are installed within buildings. This covers heating, ventilation, and air-conditioning systems that are incorporated into buildings (Figure 1-4).

International Fuel Gas Code

The *International Fuel Gas Code* (IFGC) has been developed to fill the void in the IMC for fuel gas appliances. It regulates the design, installation, maintenance, alteration, and inspection of appliances that utilize natural gas and liquefied petroleum (LP) gas, gaseous hydrogen systems, and related accessories.

International Plumbing Code

The *International Plumbing Code* (IPC) was created to provide regulations for the design, installation, alteration, and maintenance of the plumbing systems. The IPC governs the materials, sizing, and installation of potable water supply and distribution plumbing fixtures, drain waste and vent (DWV) piping, and storm drainage systems (see Figure 1-5).

International Energy Conservation Code

The *International Energy Conservation Code* (IECC) has been developed to regulate the energy use in buildings. This code provides requirements for insulation R-values and door and window insulation requirements, as well as air infiltration limitations. The IECC applies to all types of buildings that are either heated or cooled.

International Fire Code

The *International Fire Code* (IFC) has been developed for providing a reasonable level of life safety and property protection from the hazards of

FIGURE 1-4 Mechanical ductwork covered under the IMC

FIGURE 1-5 Air test on non-plastic water piping system regulated by the IPC

fire, explosion, or dangerous conditions in new and existing buildings and structures. It also provides regulations for the safety of firefighters and emergency responders during emergency operations.

International Property Maintenance Code

The *International Property Maintenance Code* (IPMC) has been developed to provide jurisdictions with a model code for the maintenance of existing buildings and properties in their jurisdictions. It provides minimum requirements and standards for premises, structures, equipment, and facilities for light, ventilation, space, heating, sanitation, protection from the elements, life safety, and safety from fire and other hazards and for safe and sanitary maintenance.

FIGURE 1-6 Single-family dwellings must comply with the requirements of the IRC

International Existing Building Code

The *International Existing Building Code* (IEBC) covers the alteration, addition, repair, relocation, and change of use to an existing building. This code provides an additional option to the IBC for the regulation of existing buildings.

International Residential Code

The *International Residential Code* (IRC) regulates the construction of one- and two-family dwellings, as well as townhouse structures (Figure 1-6). As noted previously, these types of buildings are not covered in the scope of the IBC. The IRC is designed to be a completely stand-alone code for residential construction. This code combines all of the regulations for the building, energy, mechanical, fuel gas, plumbing, and electrical into one document. This provides one source of regulation for home builders and jurisdictions. Although some requirements in the IRC are similar to the other ICC codes, it is a separate stand-alone code. It incorporates more prescriptive provisions for the construction of homes. **[Ref. 101.2]**

International Green Construction Code

The *International Green Construction Code* (IgCC) is intended to safeguard the environment, public health, safety, and general welfare through the establishment of requirements to reduce the negative potential impacts and increase the positive potential impacts of the built environment on the natural environment and building occupants. The code has minimum requirements related to conservation of natural resources, materials, and energy; the employment of renewable energy technologies; indoor and outdoor air quality; and building operations and maintenance.

You Should Know

International Swimming Pool and Spa Code (ISPSC)

Another new member in the family of the I-Codes is the ISPSC, the first comprehensive swimming pool code coordinated with the current requirements in the International Codes and APSP standards. Developed with the Association of Pool & Spa Professionals (APSP), the ISPSC establishes minimum regulations for public and residential pools, spas, and hot tubs using prescriptive and performance-related provisions. The ISPSC was published by the ICC in March 2012. ●

EXISTING BUILDINGS

Existing buildings are permitted to continue without change as long as they are maintained in accordance with the code under which they were constructed. This is sometimes called *grandfathering* or a *grandfather clause* in the code. It would not be reasonable to require buildings to comply with new provisions in the codes every 3 years they are revised. The building jurisdiction does, however, have the authority to require that existing buildings comply with minimum standards for the general safety and welfare of the public and occupants of a building. For example, if a building has been damaged by a fire, the building official can require that the building be repaired, or demolished if the structure is beyond repair. **[Ref. 102.6]**

The IBC provides regulations for additions, alterations, and repairs to an existing building. Whatever work is being done must comply with the current code. In addition, an addition or alteration cannot cause the existing building to be unsafe or in violation of the code. Those parts of the building not remodeled or affected by the new work are not required to be altered to comply with code. Essentially, those portions of the building are "grandfathered" under the code (Figure 1-7). **[Ref. 3403.1]**

When the use of a building is changed from what was originally approved, it is considered a change of occupancy. When this occurs, the building is required to comply with the requirements of the current code. For example, if a building is converted from a retail business to a restaurant, the inherent hazards change and the building must be altered to comply with the requirements for a restaurant. This can be very involved and costly, so the code allows the building official to allow the change without compliance, that is, without meeting all requirements of the code if the change of use is to a less hazardous use than the existing use based on life and fire risk. If the change is from a restaurant to a retail business, the building official may consider the retail business less hazardous than the restaurant and permit the change without having to comply with all of the code requirements (Figure 1-8). **[Ref. 3406.1]**

FIGURE 1-7 Addition to existing church building

FIGURE 1-8 Existing building regulated by the IBC, Chapter 34

International Existing Building Code Option

As an alternative to complying with the existing building provisions of the IBC, the *International Existing Building Code* (IEBC) can be used as an option. Revisions to existing buildings can create unique situations in maintaining the public health, safety, and welfare of the public. The IEBC balances the existing use of a building with the need to provide a safe building in which people can live and work. Just like the IBC, existing buildings that are not altered are permitted to continue without change, unless the safety and welfare of the public are in question.

The IEBC provides three alternative compliance paths, including prescriptive, work area, and performance. The prescriptive and performance path options are identical to the requirements in Chapter 34 of the IBC. The work area option designates different levels of alterations based on the amount of work that is being conducted. Level 1 Alterations include the removal and replacement, or the covering of existing materials or equipment with new materials for the same purpose. Level 2 Alterations include the reconfiguration of a space, or the addition or elimination of a door or window. Level 3 Alterations include any work that involves more than 50 percent of the aggregate area of the building. Depending on the level of alteration, life safety aspects of the existing building may need to be brought up to current standards. The work area option also includes requirements for repairs, additions, and change of occupancies.

Using the Code

The front matter in the IBC has a comprehensive description of effectively using the code. One of the most important pieces of information is the arrangement and format of the IBC. A table has been created to summarize the different sections and the subjects they cover. The table is replicated in Table 1-1. A description of each chapter is also provided to explain what it covers.

TABLE 1-1 IBC Organization

Chapters	Subjects
1–2	Administration and definitions
3	Use and occupancy classifications
4, 31	Special requirements for specific occupancies or elements
5–6	Height and area limitations based on the type of construction
7–9	Fire-resistance and protection requirements
10	Requirements for evacuation
11	Specific requirements to allow use and access to a building for persons with disabilities
12–13, 27–30	Building systems, such as lighting, HVAC, plumbing fixtures, elevators
14–26	Structural components—performance and stability
32	Encroachment outside of property lines
33	Safeguards during construction
34	Existing building allowances
35	Referenced standards
Appendices A–M	Appendices

When using the code, there are some marks in the margin that provide the user with information regarding the section. Solid vertical lines indicate a change from the previous edition of the code. Arrows indicate that an entire section, paragraph, exception, or table has been deleted or an item in a list of items or a table has been deleted. A single asterisk (*) in the margin indicates that text or a table has been relocated from the previous code edition. A double asterisk (**) indicates that the text or table immediately following it has been relocated there from elsewhere in the code. A summary of the relocations is provided in the preface of the code.

There are six suggested steps that may be helpful in using the IBC:

1. Classify the building.
   a. Occupancy Classification: Compute the floor area and occupant load for the building or space being considered. Determine the occupancy group for which the use of the building or space most nearly resembles.
   b. Type of Construction: Determine the type of construction being considered for the building.
   c. Allowable Floor Area: Determine the allowable floor area of each story and the building based on the occupancy classification, type of construction, and the allowable area modifications for fire sprinklers and frontage.
   d. Allowable Building Height: Compute the height of the building. Determine the allowable height of the building in feet and stories based on the occupancy classification, type of construction, and permitted increases for fire sprinklers.
   e. Location on Property: Determine the location of the building on the site and clearances to lot lines and other buildings on the same lot. Determine the fire-resistance ratings of the exterior walls and opening protection based on proximity to lot lines.
2. Review the building for occupancy requirements.
   a. Determine the requirements for mixed-occupancy buildings, including whether the building is designed as a separated or nonseparated occupancy building.
   b. Review specific occupancy requirements.
3. Review type of construction requirements.
   a. Confirm the proper type of materials and fire-resistance ratings based on the type of construction
4. Review the building or space for conformance with the accessibility and means of egress requirements.
5. Review the building or space for conformance with other detailed requirements in the IBC.
6. Review the building or space for conformity with the structural engineering regulations and requirements for the construction materials.

CHAPTER 2

Legal Aspects, Permits, and Inspections

© International Code Council

The *International Building Code* (IBC) is what is called a model code, which means that it can be adopted by a governmental agency and become law. The IBC is adopted across the country by both state and local government agencies. This chapter describes the adoption process and how the code is enforced through the review of construction plans and the inspection of the work. The building department's authority and code administration responsibilities are discussed.

CODE ADOPTION

In order for a jurisdiction to enforce the IBC, it must first adopt the code as a law. Many states adopt the code through legislation, whereas others delegate that authority to a specific board or state agency. In some states, the local county, city, town, or parish may adopt the code to regulate construction in its jurisdiction. A sample ordinance is provided in the front of the IBC to assist the jurisdiction with developing the legal portion of the adoption (Figure 2-1). The adopting legislation or ordinance goes through the public hearing process required by the adopting agency. Many jurisdictions develop amendments to the code to address local conditions and needs. When dealing with a local authority, users of the codes are advised to contact the jurisdiction and review any potential amendments.

Codes versus standards

The IBC is generally a performance-oriented code. In some cases, the code references specific documents that regulate materials and methods of construction to enhance the requirements. These documents are

SAMPLE ORDINANCE FOR ADOPTION OF THE INTERNATIONAL BUILDING CODE ORDINANCE NO.______

An ordinance of the **[JURISDICTION]** adopting the 2009 edition of the *International Building Code*, regulating and governing the conditions and maintenance of all property, buildings and structures; by providing the standards for supplied utilities and facilities and other physical things and conditions essential to ensure that structures are safe, sanitary and fit for occupation and use; and the condemnation of buildings and structures unfit for human occupancy and use and the demolition of such structures in the **[JURISDICTION]**; providing for the issuance of permits and collection of fees therefor; repealing Ordinance No. ______ of the **[JURISDICTION]** and all other ordinances and parts of the ordinances in conflict therewith.

The **[GOVERNING BODY]** of the **[JURISDICTION]** does ordain as follows:

Section 1. That a certain document, three (3) copies of which are on file in the office of the **[TITLE OF JURISDICTION'S KEEPER OF RECORDS]** of **[NAME OF JURISDICTION]**, being marked and designated as the *International Building Code*, 2009 edition, including Appendix Chapters **[FILL IN THE APPENDIX CHAPTERS BEING ADOPTED]** (see International Building Code Section 101.2.1, 2009 edition), as published by the *International Code Council*, be and is hereby adopted as the Building Code of the **[JURISDICTION]**, in the State of **[STATE NAME]** for regulating and governing the conditions and maintenance of all property, buildings and structures; by providing the standards for supplied utilities and facilities and other physical things and conditions essential to ensure that structures are safe, sanitary and fit for occupation and use; and the condemnation of buildings and struc-tures unfit for human occupancy and use and the demolition of such structures as herein provided; providing for the issuance of permits and collection of fees therefor; and each and all of the regulations, provisions, penalties, conditions and terms of said Building Code on file in the office of the [JURISDICTION] are hereby referred to, adopted, and made a port hereof, as if fully set out in this ordinance, with the additions, insertions, deletions and changes, if any, prescribed in Section 2 of this ordinance.

Section 2. The following sections are hereby revised:
Section 101.1. Insert: **[NAME OF JURISDICTION]**
Section 1612.3. Insert: **[NAME OF JURISDICTION]**
Section 1612.3. Insert: **[DATE OF ISSUANCE]**
Section 3412.2. Insert: **[DATE IN ONE LOCATION]**

Section 3. That Ordinance No. ______ of **[JURISDICTION]** entitled **[FILL IN HERE THE COMPLETE TITLE OF THE ORDINANCE OR ORDINANCES IN EFFECT AT THE PRESENT TIME SO THAT THEY WILL BE REPEALED BY DEFINITE MENTION]** and all other ordinances or parts of ordinances in conflict herewith are hereby repealed.

Section 4. That if any section, subsection, sentence, clause or phrase of this ordinance is, for any reason, held to be unconstitutional, such decision shall not affect the validity of the remaining portions of this ordinance. The **[GOVERNING BODY]** hereby declares that it would have passed this ordinance, and each section, subsection, clause or phrase thereof, irrespective of the fact that any one or more sections, subsections, sentences, clauses and phrases be declared unconstitutional.

Section 5. That nothing in this ordinance or in the Building Code hereby adopted shall be construed to affect any suit or proceeding impending in any court, or any rights acquired, or liability incurred, or any cause or causes of action acquired or existing, under any act or ordinance hereby repealed as cited in Section 3 of this ordinance; nor shall any just or legal right or remedy of any character be lost, impaired or affected by thin ordinance.

Section 6. That the **[JURISDICTION'S KEFPER OF RECORDS]** is hereby ordered and directed to cause this ordinance to be published. (An additional provision may be required to direct the number of times the ordinance is to be published and to specify that it is to be in a newspaper is general circulation. Posting may also be required.)

Section 7. That this ordinance and the rules, regulations, provisions, requirements, orders and matters established and adopted hereby shall take effect and be in full force and effect **[TIME PERIOD]** from and after the date of its final passage and adoption.

FIGURE 2-1 Example of an adopting ordinance

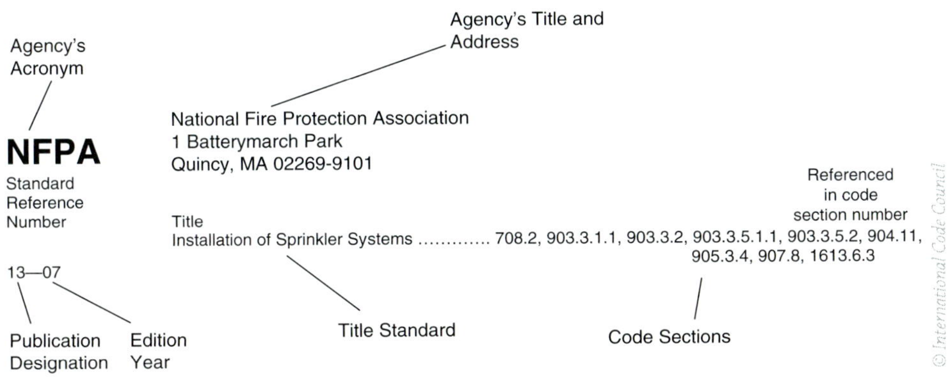

FIGURE 2-2 Referenced standard example

called *reference standards*. Reference standards are typically developed by agencies other than the International Code Council. The standards are reviewed by the code development committees to ensure that they meet ICC's requirements to be included in the code. When the standards are referenced in the code, they are considered to be a part of the code. Therefore, when a jurisdiction adopts the code, all of the standards listed in Chapter 34 are also considered to be adopted. However, only those portions of the standard that are applicable to the specific code provision apply to the specific code provision.

Many types of standards are referenced in the IBC. They include structural engineering standards, material standards, installation standards, and testing standards. The standards are referenced in different sections of the code. For example, Section 903.3.1.1 requires that when a building is required to be provided with an automatic sprinkler system in accordance with that section, the system must be installed in accordance with NFPA 13. The National Fire Protection Association (NFPA) publishes different types of fire protection standards that are referenced throughout the IBC. Basically, the IBC regulates when an automatic sprinkler system is required and NFPA 13 provides the installation requirements of the system.

Standards used in the International Building Code

Chapter 35 of the IBC lists all of the referenced standards. They are organized by the promulgating agency and initials. Each standard is listed with the document's publication designation, edition year, title, and the section of the code where the document is referenced (see Figure 2-2). It is important that the correct edition of the standard is used because the revision dates of the standards do not always coincide with the revision of the IBC. The referenced codes and standards are considered part of the requirements of the IBC. Where there are conflicts between the code provisions and the referenced document, the provisions of the code will apply.

AUTHORITY

When the IBC is adopted by a jurisdiction, it needs to be enforced. To provide for this enforcement, the jurisdiction creates the Department of Building Safety. This department is managed by the building official.

This person is appointed by the chief appointing authority of the jurisdiction. This can range from a city/county manager to the city council to the governor. The building official has the authority to enforce and interpret the code. This person also appoints deputies—building inspectors, plans examiners, and permit technicians—to assist in the enforcement of the code. Building departments range from a one-person office to hundreds of employees in an agency. **[Ref. 103]**

Because the building code is adopted as a law in a jurisdiction, it must be enforced just like any law. A department enforces the code by reviewing building plans prior to construction and issuing building permits authorizing the work. The project is then inspected by building inspectors trained and certified in their area of expertise. Inspections are also conducted by experts in specialized subjects. The inspection process is discussed further in this chapter. After the inspections are complete, a Certificate of Occupancy is issued, permitting the building to be occupied by the public (Figure 2-3).

Even though the code is adopted as a law, there are many provisions that are not designed to be specific. The building official, who has the authority to interpret the code, must understand the underlying intent and

City of Lakeside

Certificate of Occupancy

This certificate, issued pursuant to the requirements of Section 111 of the 2009 International Building Code certifies that at the time of issuance of this certificate, the structure noted below has been inspected for compliance with the applicable construction codes for the occupancy and division of occupancy and the use for which the proposed occupancy is classified.

Building Permit Number: 09-52 Building Address: 123 Main Street, Lakeside, USA

Owner Name/Address: John Doe, 234 Oak Street, Anytown, USA

Description of Work: New Office Building

Occupancy Classification: B Type of Construction: VA Design Occupant Load 100

Sprinkler System? (yes) ____ (no) X

Special Stipulations/Conditions: None

Stephen L. Thomas, CBO
Building Official

Mayor of Lakeside

Date: 1/1/12

FIGURE 2-3 Certificate of occupancy

purpose of the code provisions to provide a reasonable interpretation. The interpretation cannot have the effect of waiving the code requirements. **[Ref. 103.3]**

Several resources are available to the users of the IBC to obtain the intent of a specific code section. The most common document is the *International Building Code and Commentary* published by ICC. This document contains each section of the code followed by an explanation and description of the intent of the requirements. ICC also has a Handbook Series for the IBC available that also provides an explanation for many sections of the code. Members of ICC can also call the Council and talk with a staff engineer for assistance in applying the code. Users of the code should also contact their local jurisdiction to get the local building official's interpretation. Building departments are always willing to assist a designer or contractor in using the code. It is important to remember that the authority having jurisdiction is responsible for interpreting the code. Therefore, that authority is the sole provider of the interpretation of the code provisions.

As technology improves and new products are developed, the building code falls behind because it is only updated every 3 years. Therefore, the code gives the jurisdiction the authority to approve alternate materials, designs, and methods of construction. Although many parts of the code are prescriptive in nature, it allows a performance-based process in complying with its requirements. A *prescriptive code* is a requirement that must be specifically met. A *performance code* requirement describes the intent of a section and allows the architect to come up with a design to comply with the code.

Alternate materials and methods of construction

The building official reviews alternate designs and products to determine if they comply with the purpose and intent of the code. The official must determine if the alternate is at least the equivalent to the code in quality, strength, effectiveness, fire resistance, durability, and safety. This is accomplished by reviewing testing information and research reports provided by the manufacturer of the product. To assist the building official in this process, the ICC Evaluation Service (ICC-ES) was created. This organization develops acceptance criteria for different types of construction products. It then evaluates the products to the standard acceptance criteria to determine if the product complies with the intent of the particular code.

An evaluation report includes specific information about the product being reviewed (Figure 2-4). It provides manufacturer information, installation requirements and details, and specific conditions for the approval. The report is issued for a specific time period and renewed as required. It is also approved based on specific editions of the code. Because the code is revised every 3 years, it is important for the current report to be used in evaluating the product for use in a project. **[Ref. 104.11]**

You Should Know

Prescriptive versus Performance

Prescriptive code example

"Handrail height, measured above stair tread nosings, or finish surface of ramp slope shall be uniform, not less than 34 inches (864 mm) and not more than 38 inches (965 mm)."

Performance code example

"Fire walls shall have sufficient structural stability under fire conditions to allow collapse of construction on either side without collapse of the wall for the duration of time indicated by the required fire-resistance rating." •

ICC EVALUATION SERVICE

Most Widely Accepted and Trusted

ICC-ES Evaluation Report

ESR-4802

Issued March 1, 2011

This report is subject to re-examination in one year.

www.icc-es.org | (800) 423-6587 | (562) 699-0543 *A Subsidiary of the International Code Council®*

DIVISION: 07-THERMAL AND MOISTURE PROTECTION
Section: 07410-Metal Roof and Wall Panels

REPORT HOLDER:

ACME CUSTOM·BILT PANELS
52380 FLOWER STREET
CHICO, MONTANA 43820
(808) 664·1512
www.customblltpanels.com

EVALUATION SUBJECT:

CUSTOM·BILT STANDING SEAM METAL ROOF PANELS: CB·150

1.0 **EVALUATION SCOPE**

Compliance with the following codes:

- 2006 *International Building Code®* (IBC)
- 2006 *International Residential Code®* (IRC)

Properties evaluated:

- Weather resistance
- Fire classification
- Wind uplift resistance

2.0 USES

Custom-Bllt Standing Seam Metal Roof Panels are steel panels complying with IBC Section 1507.4 and IRC Section R905.10. The panels are recognized for use as Class A roof coverings when Installed In accordance with this report.

3.0 DESCRIPTION

3.1 Roofing Panels:

Custom-Bllt standing seam roof panels are fabricated in steel and are available In the CB-150 and SL·1750 profiles. The panels are roll-formed at the jobsite to provide the standing seams between panels. See Figures 1 and 3 for panel profiles.

The standing seam roof panels are roll-formed from minimum No. 24 gage [0.024 Inch thick (0.61 mm») cold-formed sheet steel. The steel conforms to ASTM A 792, with an aluminum-zinc alloy coating designation of AZ50.

3.2 Decking:

Solid or closely fitted decking must be minimum $^{15}/_{32}$-Inch-thlck (11 .9 mm) wood structural panel or lumber sheathing, complying with IBC Section 2304.7.2 or IRC Section R803, as applicable.

4.0 INSTALLATION

4.1 General:

Installation of the Custom-Blit Standing Seam Roof Panels must be In accordance with this report, Section 1507.4 of the IBC or Section R905.10 of the IRC, and the manufacturer's published Installation Instructions. The manufacturer's installation Instructions must be available at the Job site at all times during Installation.

The roof panels must be installed on solid or closely fitted decking, as specified In Section 3.2. Accessories such as gutters, drip angles, fascias, ridge caps, window or gable trim, valley and hip flashings, etc., are fabricated to suit each job condition. Details must be submitted to the code official for each Installation.

4.2 Roof Panel Installation:

CB-150: The CB-150 roof panels are installed on roofs having a minimum slope of 2:12 (17 percent). The roof panels are installed over the optional underlayment and secured to the sheathing with the panel clip. The clips are located at each panel rib side lap spaced 6 inches (152 mm) from all ends and at a maximum of 4 feet (1.22 mm) on center along the length of the rib, and fastened with a minimum of two No. 10 by 1-inch pan head corrosion-resistant screws. The panel ribs are mechanically seamed twice, each pass at 90 degrees, resulting in a double locking fold.

4.3 Fire Classification:

The steel panels are considered Class A roof coverings In accordance with the exception to IBC Section 1505.2 and IRC Section R902.1.

4.4 Wind Uplift Resistance:

The systems described In Section 3.0 and Installed in accordance with Sections 4.1 and 4.2 have an allowable wind uplift resistance of 45 pounds per square foot (2.15 kPa).

5.0 CONDITIONS OF USE

The standing seam metal roof panels described In this report comply with, or are suitable alternatives to what Is specified in, those codes listed In Section 1.0 of this report, subject to the following conditions:

5.1 Installation must comply with this report, the applicable code, and the manufacturer's published Installation instructions. If there is a conflict between

ICC-ES Evaluation Reports are not to be construed as representing aesthetics or any other attributes not specifically addressed, nor are they to be construed as an endorsement of the subject of the report or a recommendation for its use. There is no warranty by ICC Evaluation Service, LLC, express or implied, as to any finding or other matter in this report, or as to any product covered by the report.

FIGURE 2-4 ICC Evaluation report

PERMITS

Building permits are issued after the building department reviews the building plans for compliance with the applicable codes. When a property owner wants to build a new building, remodel a building, or build an addition onto an existing building, he or she must obtain a building permit from the local jurisdiction. A set of plans drawn by an architect or engineer is submitted to the building department, showing the type of work that will be done on the project.

The building department then reviews the drawings to determine whether or not the plans comply with the adopted codes in the jurisdiction. Once it is determined that the drawings comply with the minimum codes, a building permit is issued to the owner or contractor. The contractor is then authorized to start the work outlined in the plans. The contractor is required to keep the permit on the job site for the inspectors as the job progresses.

It is important that owners of a building, as well as the architect or contractor, obtain a building permit from their local building department. The building permit ensures that the work is completed properly and the building is constructed to the minimum standards of the code. Occupants of a building expect a building to be safe when they enter. Working on a building without a permit and assistance from the building department may create an unsafe condition for the occupants of the building. If someone does construction without the proper permits, that individual may be subject to criminal prosecution and the work may need to be removed. The individual also risks the chance of additional liability if someone is injured in the building and it is determined that a building permit and the required inspections were not obtained.

Not every little bit of work in a building requires a building permit. The code exempts small projects like storage sheds less than 120 square feet, fences over 7 feet, retaining walls less than 4 feet, painting, wallpaper installations, playground equipment accessory to one- and two-family dwellings (Figure 2-5), and other minor construction. It also allows owners of buildings to do minor maintenance without needing a building permit. Although a permit may not be required, the work that is done must comply with the code. No inspections are required on this work, but the owner is obligated to do the work properly.

FIGURE 2-5 Playground equipment accessory to a single-family dwelling does not require a permit

Submittal requirements

To obtain a permit from a jurisdiction, an applicant must submit a set of drawings and specifications showing what type of work will be done. These documents must also show how the project will comply with the applicable portions of the code. A typical set of drawings includes floor plans, elevations, sections, details, and specifications on materials and installation of equipment. The drawings should give enough information that anyone could take the plans and build the structure without knowing anything about the project. The code requires that a minimum of two sets of documents be submitted. Once a permit is issued, one set is kept on the job site and the other is kept on file in the building department. **[Ref. 107.2]**

The code provides specific information that is required to be submitted on the construction documents. The plans must show the location, construction, size, and character of the means of egress. This includes the number of occupants that will be in the building. The design of the exterior wall envelope must also be shown on the drawings. Details on the flashing, water-resistive membrane, and means of drainage must be included. Manufacturer's installation instructions are needed to ensure that penetrations and opening details are done correctly. A site plan showing the location of the building on a lot, as well as the dimensions between the building and the lot lines, must be provided (Figure 2-6). When the building is located in a flood hazard area, the plans must indicate the location of the flood hazard areas and the elevation of the 100-year flood. The most common way of showing this information on the drawings is with a Code Data Sheet. This sheet provides an overview of the code requirements for the building.

In most states, construction documents are required to be prepared by a licensed architect and/or engineer. The IBC requires an owner to engage and designate a registered design professional to be in charge of the project. If the owner changes the design professional, he or she must

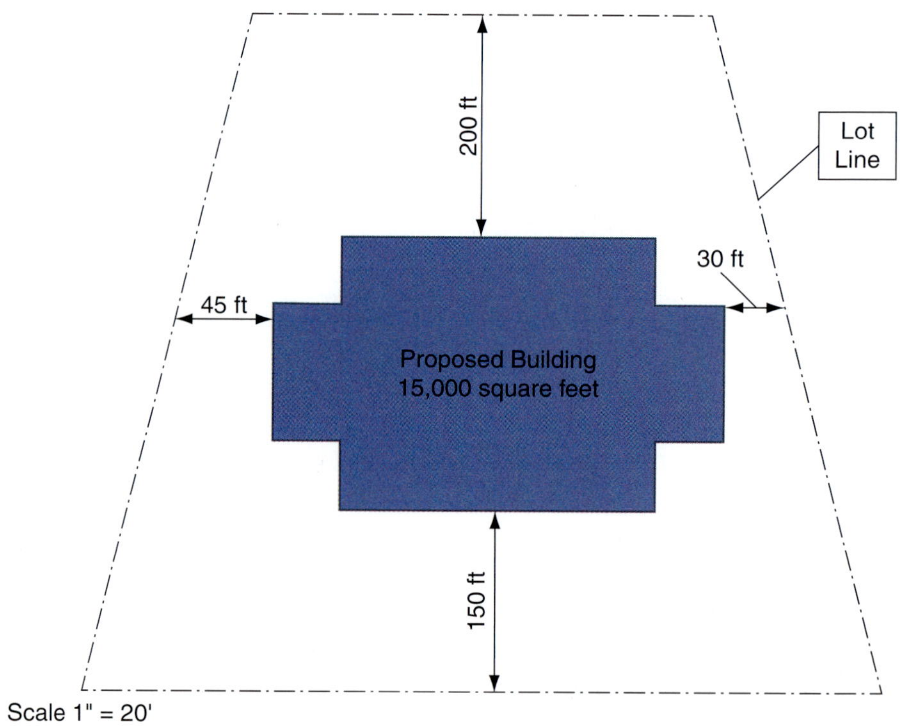

FIGURE 2-6 Site plan example

notify the building official in writing. The registered design professional is responsible for reviewing and coordinating any documents submitted by other people involved in the project and must ensure that the information is compatible with the design of the building. [Ref. 107.4.3]

Plan review process

When construction documents are submitted to the building department, the code requires that the plans be reviewed for compliance. It also requires that the review include other applicable laws of the jurisdiction. If the plans are found to be in compliance with the applicable codes, a permit can be issued for the work. If the plans have discrepancies and do not comply with the codes, a plan review report is generated and provided to the applicant. The applicant then makes corrections to the plans and resubmits them to the building department for review and a permit.

After the permit is issued, any changes that are made to the plans must be submitted for review. This includes changes that are made in the field. In some cases, information is not available when the original construction documents are submitted. Therefore, the code allows this information to be provided as *deferred submittals*. The use of deferred submittals must be approved by the building official at the time of the original submittal. The registered design professional must review the deferred submittals prior to submitting them to the building official.

INSPECTIONS

As work progresses on a project that has been issued a building permit, the building department conducts inspections to confirm compliance with the plans and the IBC. The owner or contractor is responsible for contacting the building department and requesting the appropriate inspection. Work is not permitted to be covered until an inspection has been completed. The code requires the following specific inspections be done to ensure the building complies with the applicable provisions of the code. [Ref. 110.3]

Footing and foundation inspection

The footings and foundation comprise the supporting structure bearing on the ground. The footing inspection is done after the forms and reinforcing steel are in place. The inspector checks the forms to confirm that they are the correct size as shown on the plans and also confirms that the reinforcing steel is the correct size and placed in the proper location. Foundation walls can be constructed with many different types of materials. A concrete foundation is inspected after the forms have been constructed and the reinforcement steel is installed (Figure 2-7). The inspector once again confirms

FIGURE 2-7 Concrete foundation

FIGURE 2-8 Underground plumbing must be inspected prior to replacing concrete

FIGURE 2-9 Cold-formed steel light-frame construction framing

that the installation complies with the code and plans. Masonry foundations are checked for the correct masonry, mortar, and grout, as well as the reinforcing steel. **[Ref. 110.3.1]**

Concrete slab and under-floor inspection

Many concrete slabs contain reinforcement steel, conduits, piping, and other equipment installed under and within the concrete. After the concrete is placed, it is impossible to confirm that the materials were installed properly. Therefore, the inspector must check to see that they are installed properly. This inspection is typically done after the plumbing inspector checks the plumbing installation and the electrical inspector checks the underground electrical equipment (Figure 2-8). **[Ref. 110.3.2]**

Lowest floor elevation

In areas where flooding is possible, it is important that the building be elevated above the 100-year flood level. This required inspection confirms that the lowest floor that will be occupied will be elevated at or above flood elevation. **[Ref. 110.3.3]**

Frame inspection

It is important for the structure of the building to be checked against the plans and the code. Therefore, the code requires that the structure or frame of the building be inspected. An inspector confirms that the size and installation of the structural members in the building are correct. A frame inspection is done after all of the plumbing, mechanical, and electrical systems are installed, inspected, and approved (Figure 2-9). **[Ref. 110.3.4]**

Lath and gypsum board inspection

Many buildings use gypsum board (drywall) to provide fire protection of the building. The proper installation of these materials must comply with specific installation standards and methods. Therefore, the IBC requires that the inspector confirm that the material be installed in such a manner that will provided the type of protection needed in the building. **[Ref. 110.3.5]**

Fire-resistant penetrations

Penetrations of fire-rated walls, floors, and ceilings are holes in the fire-resistive membrane (Figure 2-10). The IBC requires that these holes be protected with a material that fills them when the pipe or wire melts out in a fire. The materials used to provide this protection must be installed in a very specific way. The code requires that these systems be inspected to show that they are installed in accordance with the manufacturer's installation instructions. [Ref. 110.3.6]

FIGURE 2-10 Through-penetration fire-stop system

Energy efficiency inspections

Energy efficiency in buildings has become more important in the codes over the past several years. Therefore, the IBC requires that the materials installed in buildings for energy efficiency be inspected to ensure that they are installed correctly (Figure 2-11). The inspector checks for the proper insulation values, the proper sealing of the building, and proper equipment installation. [Ref. 110.3.7]

FIGURE 2-11 Roof insulation must be inspected prior to covering

Final inspection

The final inspection is done when the building is ready for occupancy. The inspector confirms that all of the work covered by the building permit has been completed according to the plans and applicable code. When the final inspection is approved, the owner receives a Certificate of Occupancy, which allows legal use of the building. [Ref. 110.3.10]

Third-party inspections

Many parts of a building are complex in nature and design. The typical building department may not have the necessary resources or knowledge to inspect these complex issues. Therefore, the IBC requires that certain portions of a building be inspected by a third-party inspection company that has the expertise in these types of materials and designs. The code prescribes the type and frequency of the inspections based on the type of material being installed and the method used to construct the building. The third-party inspector is hired by the building owner and is responsible for reporting his or her findings to the building department.

Third-party inspections are required on materials like structural concrete, steel, and masonry. For example, it is important to confirm that the concrete in a foundation provides the required strength to support a building. Samples of the concrete are taken during the placement of the concrete and test cylinders are created and broken to determine the strength of the concrete. Other tests are done on the concrete to determine its consistency and compliance with the project specifications. [Ref. 1704]

BOARD OF APPEALS

The building official is responsible for the interpretation of the regulations in the IBC. Occasionally, an architect or contractor may disagree with the interpretation provided by the building official. In this case, the architect or contractor can request a hearing before the jurisdiction's Board of Appeals. The Board of Appeals is made up of people appointed by the local authority. Typically, the people on the board are familiar with construction and understand the building process. **[Ref. 113]**

Members of the Board of Appeals listen to both sides of an issue that comes before them. They review the information against the intent and purpose of the code and render a decision regarding the interpretation of a code requirement. The board's position is then carried out by either the building department or contractor.

DEFINITIONS

Many terms in the IBC are specific to the code. Therefore, the code includes definitions for different terms in the regulations. When a term is not specified in the code, the normal use of the term can be used as it applies to the code. The definitions listed in the IBC are very important to the use of the code. Defined terms are italicized in the code language. When a user sees these terms, they should reference Chapter 2 of the IBC to determine the specific use of the term in the code. It provides important clarification to the other sections of the code. All of the definitions are contained in Chapter 2 of the IBC. Cross references in specific chapters are provided in the first part of many chapters. Some of the IBC-specific definitions have been provided in the glossary of this book. **[Ref. 201]**

PART

II

Building Planning

© iStockphoto/Yakobchuk

CHAPTER

3 Building Use

© International Code Council

Buildings are used in different ways. There are different hazards in each type of building. Some buildings may have very limited hazards and others may have very serious hazards. The IBC divides the uses into different basic occupancy classifications based on the hazards in the building. There are ten different occupancy classifications in the code. The inherent hazards in buildings can create different problems for the occupants. They may create a problem with providing a safe way to get out of a building or there may be materials used or stored in the building that are dangerous to the occupants. It is important to determine the occupancy classification of the building right away. Most of the requirements in the IBC are based on the occupancy classification of the building. Without an occupancy classification, the code cannot be used.

The IBC provides regulations on the design of the building based on these inherent conditions and hazards. It regulates the types of materials used in the construction of the building. It regulates the size and height of the building based on the use and the type of materials used in the construction. The exit, or means of egress, system in the building is designed based on the hazards in the building. Fire protection systems are also based on the use of the building.

BASIC OCCUPANCIES

As noted in the introductory paragraph, there are ten different basic *occupancies* in the IBC. Many of those occupancies are then divided into separate subcategories (see Table 3-1). The majority of the subcategories are generally listed from the most hazardous use to the least hazardous within the major occupancy classification. For example, a Group F-1 occupancy housing a moderate-hazard factory has more potential hazards than a Group F-2 occupancy that contains a low-hazard factory. It can be argued though that a Group I-1 occupancy is not the most hazardous occupancy in the Institution category. To assist in determining the classification of a building, many of the sections outlining the occupancies have a list of examples. Many buildings have multiple occupancies in them. When this happens, each individual area must be classified into the appropriate occupancy classification. Chapter 6 discusses how these multioccupancy buildings are addressed. **[Ref. 302]**

TABLE 3-1 Occupancy classifications

Occupancy Group	Description
A-1	Assembly uses, usually with fixed seating, intended for the production and viewing of the performing arts or motion pictures
A-2	Assembly uses intended for food and/or drink consumption
A-3	Assembly uses intended for worship, recreation, or amusement and other assembly uses not classified elsewhere in Group A
A-4	Assembly uses intended for viewing of indoor sporting events and activities with spectator seating
A-5	Assembly uses intended for participation in or viewing outdoor activities
B	The use of a building or structure, or a portion thereof, for office, professional, or service-type transactions, including storage of records and accounts.
E	The use of a building or structure, or a portion thereof, by six or more persons at any one time for educational purposes through the 12th grade and child care facilities.
F-1	Factory industrial uses which are not classified as Factory Industrial F-2 Low Hazard
F-2	Factory industrial uses that involve the fabrication or manufacturing of noncombustible materials which during finishing, packing, or processing do not involve a significant fire hazard
H-1	Buildings and structures containing materials that pose a detonation hazard
H-2	Buildings and structures containing materials that pose a deflagration hazard or a hazard from accelerated burning
H-3	Buildings and structures containing materials that readily support combustion or that pose a physical hazard
H-4	Buildings and structures which contain materials that are health hazards

(continues)

TABLE 3-1 Occupancy classifications (Continued)

Occupancy Group	Description
H-5	Semiconductor fabrication facilities and comparable research and development areas in which hazardous production materials (HPM) are used.
I-1	Buildings, structures, or parts thereof housing more than 16 persons, on a 24-hour basis, who because of age, mental disability, or other reasons, live in a supervised residential environment that provides personal care services. The occupants are capable of responding to an emergency situation without physical assistance from staff.
I-2	Buildings and structures used for medical, surgical, psychiatric, nursing, or custodial care on a 24-hour basis for more than five persons who are not capable of self-preservation.
I-3	Buildings and structures that are inhabited by more than five persons who are under restraint or security. An I-3 facility is occupied by persons who are generally incapable of self-preservation due to security measures not under the occupants' control.
I-4	Buildings and structures occupied by persons of any age who receive custodial care for less than 24 hours by individuals other than parents or guardians, relatives by blood, marriage, or adoption, and in a place other than the home of the person cared for.
M	Buildings and structures or a portion thereof, for the display and sale of merchandise, and involves stocks of goods, wares, or merchandise incidental to such purposes and accessible to the public.
R-1	Residential occupancies containing sleeping units where the occupants are primarily transient in nature
R-2	Residential occupancies containing sleeping units or more than two dwelling units where the occupants are primarily permanent in nature
R-3	Residential occupancies where the occupants are primarily permanent in nature and not classified as Group R-1, R-2, R-4, or I
R-4	Buildings arranged for occupancy as residential care/assisted living facilities including more than five but not more than 16 occupants, excluding staff.
S-1	Buildings occupied for storage uses that are not classified as Group S-2
S-2	Buildings used for the storage of noncombustible materials such as products on wood pallets or in paper cartons with or without single thickness divisions; or in paper wrappings. Such products are permitted to have a negligible amount of plastic trim, such as knobs, handles, or film wrapping.
U	Buildings and structures of an accessory character and miscellaneous structures not classified in any specific occupancy shall be constructed, equipped, and maintained to conform to the requirements of this code commensurate with the fire and life hazard incidental to their occupancy.

FIGURE 3-1 A place of worship is classified as a Group A-3 occupancy

Assembly occupancies

When a building is used by a large number of people for the purpose of assembling, it is classified as a Group A occupancy. People tend to panic when a fire or some other type of dangerous event occurs. When there are a large number of people in the building and these events occur, it is very important to protect them and provide a safe and efficient way for them to get out of the building. Assembly occupancies include movie theaters, restaurants, nightclubs, places of worship, community and recreation centers, sports arenas, and stadiums (Figure 3-1). **[Ref. 303]**

Many buildings have rooms for small-assembly uses such as conference rooms and small break rooms. If the room or space is less than 750 square feet in area or it has an occupant load less than fifty and it is accessory to a different occupancy, the room is classified as a Group B occupancy or as part of the main occupancy. Buildings or tenant spaces such as small restaurants with an occupant load less than fifty are also classified as Group B occupancies. **[Ref. 303.1]**

Business occupancies

Buildings or spaces that include personal service operations and offices are classified as Group B occupancies. These occupancies may be the safest occupancies in the code. People that are in Group B occupancies are generally alert and aware of their surroundings. The number of people is relatively low compared to Group A occupancy. If a fire or some other type of event occurs, the occupants can usually reach a safe place in a reasonable time and effort. Examples of Group B occupancies include professional offices, banks, post offices, and colleges. **[Ref. 304]**

Educational occupancies

Schools providing education for children up to and including those in the twelfth grade are called Group E occupancies. This classification also includes child care facilities for children older than 2½ years of age. These occupancies have inherent problems that limit the level of safety for the users of the building. The first problem is the age of the occupants. Children require direction to evacuate a building when a catastrophic event occurs. Many schools practice this during their scheduled fire drills. Examples of Group E occupancies include day care facilities and elementary, middle, and high schools (Figure 3-2). **[Ref. 305]**

Many schools have gymnasiums and theaters associated with the occupancy. These types of uses would normally be classified as a Group A occupancy. However, the code considers these areas to an extension of the educational function. Therefore, these assembly areas are permitted to be classified as a Group E occupancy with the rest of the building. The means of egress is still required to be designed as if it is an assembly occupancy though. **[Ref. 303.1.3]**

FIGURE 3-2 An elementary school is classified as a Group E occupancy

Factory/Industrial occupancies

Buildings used to assemble, manufacture, package, or process products are classified as Group F occupancies. Group F occupancies are separated into two very different subcategories based on the combustibility of the materials that are manufactured in the building. These uses may have inherent risks due to the materials used, as well as the overall large size of these uses. Factory/industrial occupancies include factories, bakeries, book publishers, woodworking establishments, and foundries. [Ref. 306]

Hazardous occupancies

When a building contains reasonably large amounts of hazardous materials, it falls within the Group H occupancy classification. Hazardous materials are classified as either physical or health hazards. Physical hazard materials are materials that burn, have accelerated burning, or produce an explosion. Health hazard materials can cause death or injury or can incapacitate a person after a single, brief exposure. These occupancies are obviously in the most hazardous group in the IBC. Some hazardous materials are permitted in other occupancy groups as long as they do not exceed a specific amount. However, when those amounts exceed a maximum allowable quantity of materials, the building receives the hazardous occupancy classification. There are five different levels of hazardous occupancies based on the level of hazard. This occupancy includes buildings where hazardous materials are stored, used, or manufactured. [Ref. 307]

Institutional occupancies

Whenever a building houses people who cannot help themselves in case of an emergency, it is classified as a Group I occupancy. This classification has four subcategories of uses. This group includes assisted living facilities in which people are older or disabled and may need some direction to egress but are capable of getting out on their own. It also includes hospitals in which people may be unconscious or sick or injured enough that they cannot get out of a building on their own. Jails and prisons are also included in institutional occupancies (Figure 3-3). The last type of Group I occupancies are adult care facilities and child care facilities for children less than 2½ years of age. The classification is based on the amount of time that occupants spend in the building and the level of assistance they may need to respond to an emergency. The classification is also based on the number of people that are housed in the building. If the building houses less than sixteen persons, it would be considered a Residential Occupancy. [Ref. 308]

FIGURE 3-3 A county jail falls within the Group I-3 occupancy

Mercantile occupancies

Group M occupancies include uses wherein merchandise is displayed and sold to the public. These occupancies include retail stores, wholesale stores, shopping centers, and gas stations where no vehicle repairs are conducted. Group M occupancies are relatively safe depending on the types of products being sold. People are generally alert and aware of their surroundings and, therefore, can egress the building on their own. [Ref. 309]

Residential occupancies

Buildings in which people live and sleep are designated as Group R occupancies. There are four subcategories of Group R occupancies, depending on the number of dwelling units in a building and whether or not people are familiar with their surroundings. Any time people are asleep in a building, their reaction time to an emergency is delayed. There are also inherent hazards when people are cooking or just living in a dwelling unit. This occupancy group also includes buildings that have one or two housing units and cannot be constructed under the *International Residential Code* (Figure 3-4). **[Ref. 310]**

Storage occupancies

When a building is used to store materials or products that do not exceed the hazardous materials limits outlined in previous occupancy descriptions, it is classified as a Group S occupancy. There are two separate subcategories of Group S occupancies depending on the combustibility of the materials being stored, similar to Group F, factory/industrial occupancies. Examples of storage occupancies include warehouses, motor vehicle repair garages, and automobile parking garages. **[Ref. 311]**

Utility occupancies

Group U occupancies include buildings or areas that are accessory in nature to a main occupancy. These buildings are generally low-hazard occupancies. They include miscellaneous structures such as fences and retaining walls. Group U occupancies include private garages, barns, sheds, and agricultural buildings (Figure 3-5). **[Ref. 312]**

FIGURE 3-4 An apartment building is classified as a Group R-2 occupancy

FIGURE 3-5 A private garage is classified as a Group U occupancy

CHAPTER

4 Construction Types

© International Code Council

The type of materials and the fire protection of those materials play a large role in the size of a building. The IBC classifies structures based on these factors for determining the maximum height and floor area of the buildings. This classification is called the *type of construction*. Buildings with a low type of construction are limited in size, whereas those with a higher type of construction are allowed to be bigger and in some cases unlimited in height and floor area. The major types of construction are I, II, III, IV, and V. Each of these classifications, except Type IV, is split into subcategories of A or B (see Table 4-1). The most restrictive type of construction is Type I and the least restrictive is Type V. This chapter discusses the differences between the types of construction and how they are addressed in the design of a building.

TABLE 4-1 Fire-resistance rating requirements for building elements (hours) (IBC Table 601)

	Type I		Type II		Type III		Type IV	Type V	
Building Element	**A**	**B**	**A[d]**	**B**	**A[d]**	**B**	**HT**	**A[d]**	**B**
Primary structural frame[g] (see Section 202)	3[a]	2[a]	1	0	1	0	HT	1	0
Bearing walls									
Exterior[f,g]	3	2	1	0	2	2	2	1	0
Interior	3[a]	2[a]	1	0	1	0	1/HT	1	0
Nonbearing walls and partitions, exterior	see Table 602								
Nonbearing walls and partitions, interior[e]	0	0	0	0	0	0	See Section 602.4.6	0	0
Floor construction and secondary members (see Section 202)	2	2	1	0	1	0	HT	1	0
Roof construction and secondary members (see Section 202)	1½[b]	1[b,c]	1[b,c]	0[c]	1[b,c]	0	HT	1[b,c]	0

For SI: 1 foot = 304.8 mm.

a. Roof supports: Fire-resistance ratings of primary structural frame and bearing walls are permitted to be reduced by 1 hour where supporting a roof only.

b. Except in Groups F-1, H, M, and S-1 occupancies, fire protection of structural members shall not be required, including protection of roof framing and decking where every part of the roof construction in 20 feet or more above any floor immediately below. Fire-retardant-treated wood members shall be allowed to be used for such unprotected members.

c. In all occupancies, heavy timber shall be allowed where a 1-hour or less fire-resistance rating is required.

d. An approved automatic sprinkler system in accordance with Section 903.3.1 shall be allowed to be substituted for 1-hour fire-resistance-rated construction, provided such system is not otherwise required by other provisions of the code or used for an allowable area increase in accordance with Section 506.3 or an allowable height increase in accordance with Section 504.2. The 1-hour substitution for the fire resistance of exterior walls shall not be permitted.

e. Not less than the fire-resistance rating required by other sections of this code.

f. Not less than the fire-resistance rating based on fire separation distance (see Table 602).

g. Not less than the fire-resistance rating as referenced in Section 704.10.

COMBUSTIBLE VERSUS NONCOMBUSTIBLE

The more restrictive construction types require noncombustible materials to be used in what the code calls a *building element*. A building element is an essential part of the overall structure. It includes the actual structural portion of the building, including the columns, beams, bearing and nonbearing walls, floor construction, and roof construction. These elements are a fundamental portion of the building. Without them, the building could fail or collapse in a fire or other type of disaster. Therefore, the code requires these elements to be noncombustible in larger and taller buildings, whereas it allows any materials permitted in the code, including combustible materials, in the building elements in smaller buildings. **[Ref. 202]**

Buildings of Type I and Type II construction are required to be constructed of wholly noncombustible materials. Common noncombustible materials include concrete, steel, and masonry (Figure 4-1). Other materials that have been tested to a specific standard can also be considered noncombustible under the IBC. Type III and Type IV construction permit a mixture of combustible and noncombustible materials. The exterior walls of these structures are required to be constructed

FIGURE 4-1 Type IA construction: All building elements must be noncombustible

FIGURE 4-2 Type V construction: Any materials listed in the IBC can be used

of noncombustible materials or of lumber that has been pressure treated to reduce the combustibility. The rest of the essential components of the building are permitted to be constructed out of any materials permitted by the code including combustible materials. Type V construction allows the use of any materials, including combustible materials, throughout the structure (Figure 4-2). **[Ref. 602]**

Even though the code requires noncombustible materials in buildings of Type I and Type II construction, there are cases in which the code permits combustible materials within these buildings. The code allows these exceptions because the materials are not essential to the overall stability of the building, but there are common areas where such materials are used. Examples of areas that are permitted to have combustible materials in Type I and Type II construction include the following: **[Ref. 603]**

- Thermal and acoustical insulation, other than foam plastics
- Roof coverings
- Interior floor coverings
- Interior wall and ceiling finishes
- Decorative materials or trim
- Millwork such as doors, door frames, window sashes, and frames
- Combustible exterior wall coverings, balconies and similar projections, and bay or oriel windows
- Blocking such as for handrails, millwork, cabinets, and window and door frames
- Nailing or furring strips
- Fire-retardant treated wood in specific locations

PROTECTED VERSUS UNPROTECTED

As noted previously, many of the classifications for types of construction are divided into two subcategories, A or B. This designation identifies the overall fire protection of the building elements. As the sizes of buildings increase, it is more important to protect the structure from fire. Therefore, the IBC's Table 601 (shown in Table 4-1 of this chapter) prescribes the fire protection of the essential portions of the building.

FIGURE 4-3 Type I construction with steel protected with sprayed applied fireproofing

The numbers in the table indicate what level of fire protection is required in hours. For example, an interior bearing wall in a Type VA building is required to be protected from fire for a time period of 1 hour, whereas the same wall in a Type IA building is required to be protected for 3 hours (Figure 4-3). The method for determining this protection is discussed later.

Type II, Type III, and Type V construction are similar in how they are divided into the two subcategories. With the exception of exterior walls in Type III construction, the building elements that are required to be protected for 1 hour in the *A* subcategory and are not required to be protected for the *B* subcategory. Exterior walls in Type III construction are required to be protected for 2 hours. The fire protection for Type I buildings is higher because they are typically much larger buildings. Type IA construction requires that the supporting structure portion of the building be provided with a minimum fire protection of 3 hours. The floors are permitted to have a 2-hour protection, and the roof is permitted a 1½-hour fire protection. The less restrictive Type IB construction requires 2-hour protection for the beams, columns, bearing walls, and floor construction. The roof construction is permitted to be protected for 1 hour. **[Ref. 601]**

Type IV construction is typically called "heavy timber construction." The exterior bearing walls in this type of construction are required to be noncombustible materials with a 2-hour fire protection. The interior elements are then permitted to be solid or laminated wood without concealed construction. The fire protection in these types of buildings comes from the insulating capacity of the wood. When a fire impacts a thick piece of wood, it chars the first ½ to ¾ inches of the member and then goes out. This leaves a significant portion of the member intact, which will continue to support the structure in a fire.

To provide this level of fire protection, the wood members are required to be a minimum size in a Type IV building. Columns are required

FIGURE 4-4 **Heavy timber material used in Type V construction**

to be a minimum of 8″ × 8″ when supporting floors and a minimum of 6″ × 8″ when supporting a roof. Wood beams and girders must be a minimum of 6″ × 10″. The members supporting the roof must also be similar large solid or laminated wood beams or arches. In addition, the floors and roof must be constructed of solid materials without any concealed spaces. Interior partitions are required to be solid wood construction formed by not less than two layers of 1-inch boards, or laminated construction 4 inches thick, or have a 1-hour fire protection. **[Ref. 602.4]**

Although few buildings are constructed of Type IV construction, individual elements of heavy timber construction are used in other types of construction. Several provisions in the IBC allow heavy timber construction in lieu of fire-resistant rated construction. For example, heavy timber construction can be used in roof construction of all types of construction requiring a 1-hour fire-resistance rating if the designer chooses (Figure 4-4). **[Ref. Table 601, footnote c]**

Small, single-story commercial buildings are typically constructed of Type VB construction. Type VA construction is commonly used in apartment and hotel buildings up to three or four stories in height. Type IV construction is not currently used except in very special cases. Many existing buildings in older portions of city downtown areas are constructed of Type IV construction. Type III construction is typically used in large educational buildings. Schools are also constructed of Type IIA or Type IIB construction. The most common use of Type IIB construction is in big box retail stores, warehouses, and industrial buildings. Type IA construction is used for mid-rise buildings that are between five and twelve stories. High-rise buildings over twelve stories in height and very large low-rise buildings are constructed of Type IA construction.

CHAPTER 5

Building Size

In order to limit the potential damage in a building from fire and other hazards, the *International Building Code* limits the size of buildings. The overall height and area of a structure are based on the use of the building and the materials used in the construction of the building. It makes sense for a building that has inherent hazards to be limited in size. For example, a building that contains a hazardous use should not be located in a large building constructed of combustible materials. On the other side of the issue, a building that is constructed of noncombustible materials and protected from fire can be relatively larger. In some cases, these buildings are permitted to be unlimited in height and area. This chapter outlines the process to determine the size of buildings based on the occupancy classification and type of construction.

MAXIMUM HEIGHT

As already noted, the height of a building is limited based on the type of construction and occupancy classification. Table 503 of the IBC provides the basic limitations of building height. It outlines the maximum number of stories and height in feet of a building. For example, a Group R-2 apartment building of Type VB construction is permitted to be a maximum of two stories and 40 feet in height. However, the same building that is constructed of Type IA construction is not limited in height (see Table 5-1).

The height of the building in feet is measured from the average ground level around the exterior walls of the building to the average height of the highest roof on the building (see Figure 5-1). The height limitation in the code is also based on the number of "stories above grade plane." A *story* is defined in the code as "that portion of a building

TABLE 5-1 Allowable building heights (A portion of IBC Table 503)

Building height limitations shown in feet above grade plane: Story limitations shown as stories above grade plane

	Type of Construction								
	I		II		III		IV	V	
Group	A	B	A	B	A	B	HT	A	B
Height (ft)	UL	160	65	55	65	55	65	50	40
A-2	UL	11	3	2	3	2	3	2	1
B	UL	UL	5	3	5	3	5	3	2
M	UL	11	4	2	4	2	4	3	1
R-2	UL	11	4	4	4	4	4	3	2
S-1	UL	11	4	2	3	2	4	3	1

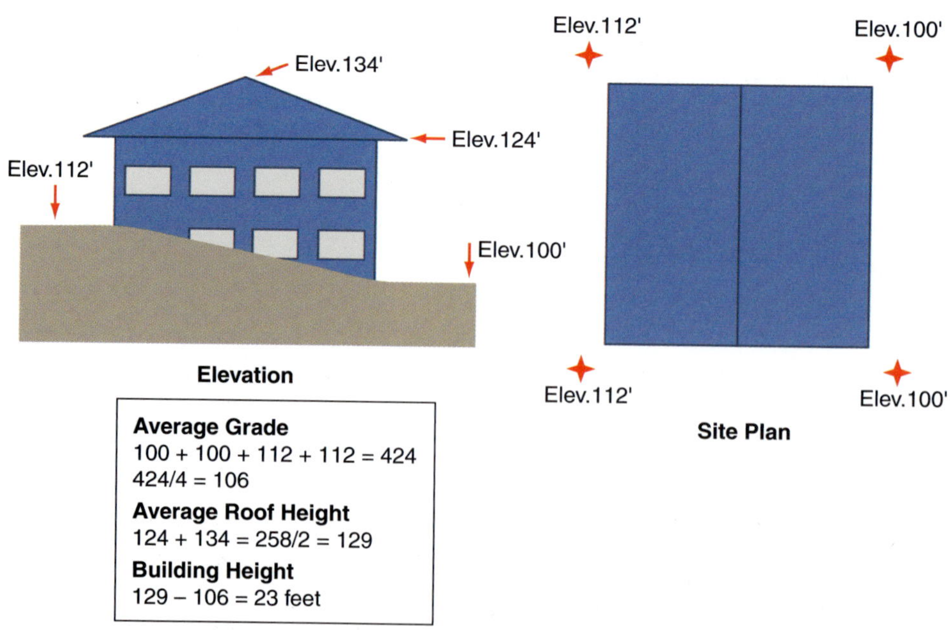

FIGURE 5-1 Building height determination

included between the upper surface of a floor and the upper surface of the floor or roof next above." Sometimes, stories occur below grade or partially below grade. The user of the code must determine if the story being evaluated is considered a "basement" or a "story above grade plane." For a story to be considered a story above grade plane, the floor above the story being assessed must be more than 6 feet above the average grade around the exterior wall of the building. In addition, if the floor above the story is more than 12 feet above grade at any point, it is considered a "story above grade plane." Any floor that is not a "story above grade plane" is considered a basement. **[Ref. 504]**

Many requirements in the code can be modified when an automatic fire sprinkler system is installed throughout a building. The fire sprinkler system must be installed in accordance with the National Fire Protection Association's Standard 13 or 13R. When either of these systems are installed throughout a building, the maximum height in feet can be increased 20 feet and one additional story. So, for the earlier example of a Group R-2 apartment building of Type VB construction, it is now permitted to be a maximum of three stories and 60 feet in height when a fire sprinkler system is installed in the building.

> **You Should Know**
>
> **Automatic Fire Sprinkler Standards**
>
> NFPA 13: Building Protection
> - Sprinklers provided in all concealed combustible spaces like attics and floor cavities
> - Sprinklers required in all rooms of a building
>
> NFPA 13R: Occupant Protection
> - Limited to residential uses with a maximum height of four stories
> - Sprinklers permitted to be omitted from concealed combustible spaces like attics and floor cavities
> - Sprinkler permitted to be omitted from small closets and restrooms
>
> NFPA 13D: Occupant Protection
> - Limited to one- and two-family dwellings ●

BUILDING AREA

The overall floor area of a building has a direct impact on several code-related concerns. Very large buildings create difficulties for firefighters because they may not be able to access portions of the building. Larger buildings also have more combustible contents that pose a problem for firefighting. People also have longer distances to travel to get out of a large building in an emergency. Therefore, the IBC requires a higher type of construction as the building gets larger. Higher types of construction require noncombustible materials and higher levels of fire protection than lower types.

The IBC starts by providing a maximum base area for a single-story building and for an individual story within a building depending on the occupancy classification and type of construction. Table 503 not only outlines the maximum height as already noted but also outlines the maximum area per story. For example, a Group R-2 apartment building of Type VB construction is permitted 7,000 square feet of floor area per story (see Table 5-2). This base area is then allowed to be increased by two factors, fire sprinklers and fire department access.

If the actual area of a single-story building is less than the values in Table 503, no further increases are necessary. For a building that is two stories in height, the value in Table 503 is multiplied by two to determine the total allowable building area. The total building area for buildings that are three or more stories in height, the values in Table 503 are multiplied by three. As long as the floor area of each of the stories is less than the values in Table 503 and the total building area is not exceeded, the building is in compliance with the IBC from the standpoint of building area (see Figure 5-2). **[Ref. 506.4]**

Fire sprinkler systems work to suppress and contain a fire, which gives the firefighters more time to respond and extinguish it. Because

TABLE 5-2 Allowable building areas (A portion of IBC Table 503)

Building area limitations shown in square feet, as determined by the definition of "Area, building" per story.

	Type of Construction								
	I		II		III		IV	V	
Group	A	B	A	B	A	B	HT	A	B
A-2	UL	UL	15,500	9,500	14,000	8,500	15,500	11,500	6,000
B	UL	UL	37,500	23,000	28,500	19,000	36,000	18,000	9,000
M	UL	UL	21,500	12,500	18,500	12,500	20,500	14,000	9,000
R-2	UL	UL	24,000	16,000	24,000	16,000	20,500	12,000	7,000
S-1	UL	48,000	26,000	17,500	26,000	17,500	25,500	14,000	9,000

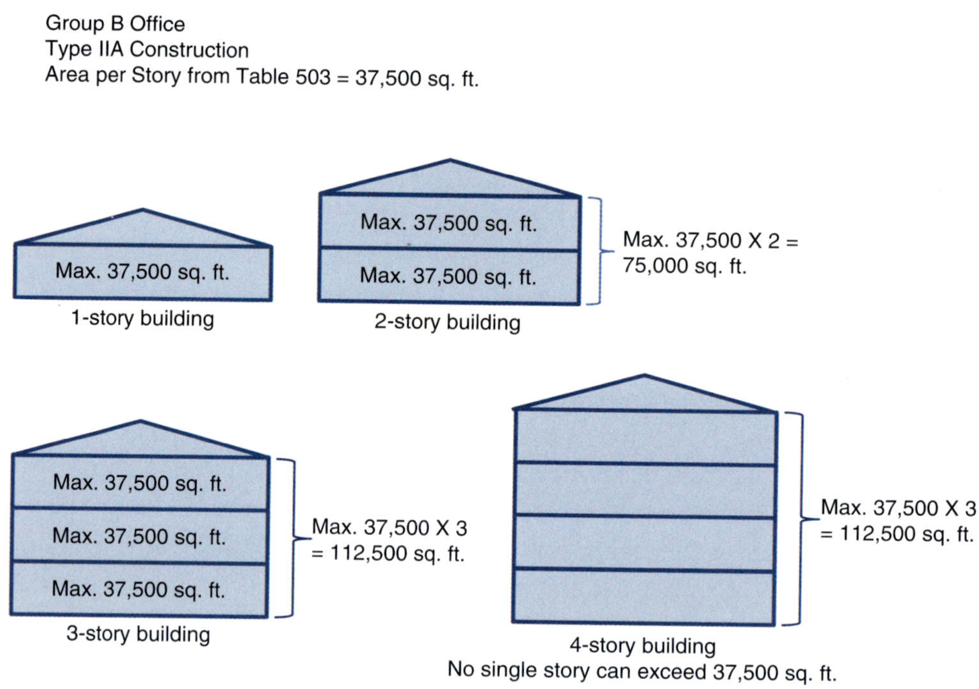

FIGURE 5-2 Maximum floor areas without increases

fire sprinklers provide this protection, a building can be increased in size. The IBC allows the designer to increase the floor area of a one-story building by 300 percent. An individual story of a multistory building can be increased 200 percent if the structure is provided with fire sprinklers throughout. The use of this increase is limited to buildings provided with an automatic sprinkler system designed in accordance with the NFPA's Standard 13. For example, a one-story Group M occupancy of Type IIB Construction has an allowable area per story of 12,500 square feet in the table. If an automatic sprinkler system is installed, the value can be increased 300 percent or 37,500 square feet. This is added to the original value for a total permitted building area of 50,000 square feet. If the same occupancy is placed in a two-story building, the single floor area value is increased 200 percent, or 25,000 square feet, which can be added to the original value for a total of 37,500 square feet of allowable area per story. Using the multi-story calculation discussed previously, the total permitted area of the two-story building would be 75,000 square feet. **[Ref. 506.3]**

In order for a fire department to fight a structure fire, it needs enough space around the building to place firefighters and equipment. The IBC uses this space around a building to increase the allowable floor area. The IBC uses the term *frontage* for this space around a building. As the amount of area or frontage around a building is increased, more floor area can be added. Designers must be careful to provide these open spaces around the building so that they are usable by the fire department. To get any credit for the frontage, the building must have a minimum of 20 feet of open frontage for at least 25 percent of the perimeter of the building. Any open space in excess of 30 feet is not given any credit for additional floor area. **[Ref. 506.2]**

To calculate the amount of floor area increase, the IBC provides an equation that is used to determine how much additional area is permitted. It determines a percentage of the exterior perimeter of the building that provides adequate fire department access. The maximum increase that a designer can use with the frontage increase is 75 percent of the basic area noted in IBC Table 503 (see Figure 5-3). The equation that is used to determine the frontage increase is as follows:

IBC Equation 5-2

$$I_f = [F/P - 0.25]W/30$$

where:

I_f = Area increase due to frontage
F = Building perimeter that fronts on a public way or open space having 20 feet minimum width
P = Perimeter of entire building (feet)
W = Width of public way or open space (feet) in accordance with Section 506.2.1

Once the frontage increase and sprinkler increase are determined, the maximum floor area per story is determined using another equation in the code. This equation takes the base area from IBC Table 503 and adds the increases outlined before. The equation used to determine the maximum area per story is as follows:

IBC Equation 5-1

$$A_a = \{A_t + [A_t \times I_f] + [A_t \times I_s]\}$$

where:

A_a = Allowable building area per story (square feet)
A_t = Tabular building area per story in accordance with Table 503 (square feet)
I_f = Area increase factor due to frontage as calculated in accordance with Section 506.2
I_s = Area increase factor due to sprinkler protection as calculated in accordance with Section 506.3

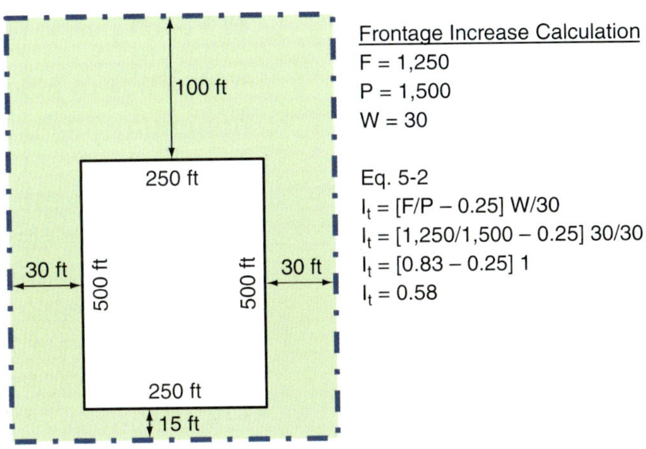

FIGURE 5-3 Frontage increase calculation

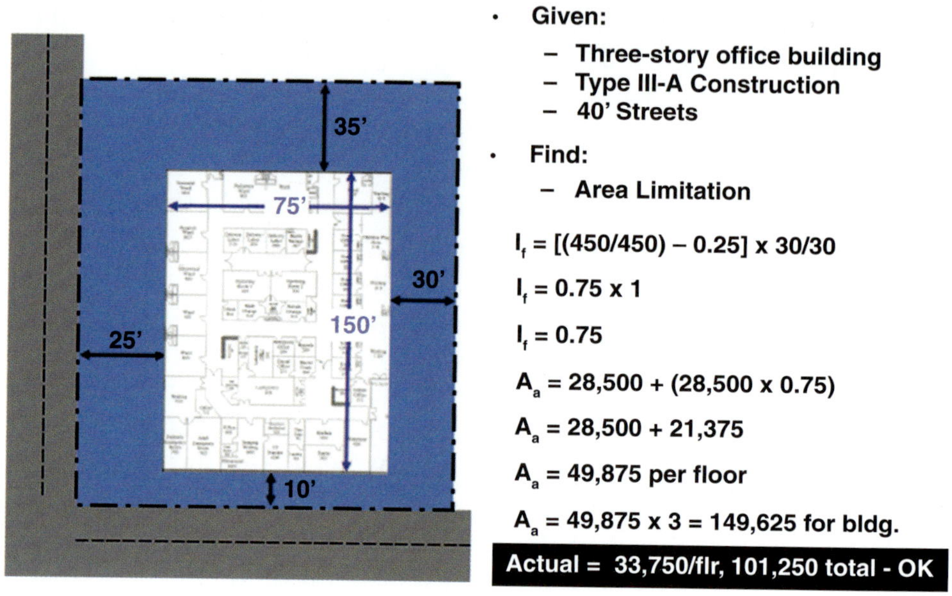

FIGURE 5-4 Allowable area calculation

Once the maximum floor area is determined for each story, the total allowable area of a building is determined by multiplying the allowable building area per story (A_a) by the number of stories in the building up to three. Buildings over three stories in height are limited to a maximum of three times the allowable area per story calculated in Equation 5-1 of the IBC (Figure 5-4).

In some cases, buildings are allowed to be unlimited in area. One option is to construct the building of Type IA construction. The IBC also allows specific uses to have unlimited areas. In order for these uses to be unlimited in area, they must meet certain criteria. For example, a single-story Group S-1 warehouse building can be any size as long as it is provided with fire sprinklers throughout and has open space or frontage a minimum of 60 feet wide. These buildings can reach very large areas. **[Ref. 507]**

INTERIOR DIMENSIONS

For people to have a habitable space in which to live, the IBC requires minimum dimensions in a building. Ceilings in habitable spaces must be a minimum of 7 feet, 6 inches to provide people enough headroom to feel comfortable. A habitable space is defined as a space in a building for living, sleeping, eating, or cooking. Bathrooms, toilet rooms, closets, halls, storage or utility spaces, and similar areas are not considered habitable spaces. The IBC also requires habitable rooms to be a minimum of 7 feet in width. This provides adequate room for furniture and space for the intended activity. The only exception to this rule is non-accessible kitchens for which a minimum clear passageway of not less than 3 feet must be provided. This provision is typically

FIGURE 5-5 Required barrier under projecting structural element (protruding object)

applied to residential occupancies and is consistent with the requirements in the *International Residential Code*. **[Ref. 1208.2]**

It is important to pay attention to objects hanging down from the ceiling or sticking out from a wall. When these objects protrude into an area in which people are walking, they can become a hazard. Therefore, the IBC requires that a minimum 80-inch clearance be provided below such protruding objects. If the clearance cannot be provided, a barrier must be installed below the protruding object. This barrier provides a visual cue for people to see that a danger exists and for people who are visually impaired to hit the barrier with their cane before they hit their head on the protruding object (Figure 5-5). **[Ref. 1003.3]**

Residential dwelling units are also required to have at least one room that has a minimum of 120 square feet. All other habitable rooms are required to be 70 square feet in area. Kitchens in dwelling units only need to be 50 square feet. One type of dwelling unit that does not have to meet these requirements is an *efficiency unit*. These units are sometimes called *studio apartments* and usually do not have separate sleeping areas. **[Ref. 1208.4]**

Specific dimensions are also required in other portions of a building. For example, the minimum ceiling height of the exit path is also required to be a minimum of 7 feet, 6 inches. Egress doors must have a minimum clear width of 32 inches and a minimum 80-inch clear height. Disabled accessibility requirements are discussed in Chapter 11. **[Ref. 1003.2 and 1008.1.1]**

Code Basics

Efficiency units must comply with the following:

1. A living room of not less than 220 square feet of floor area. An additional 100 square feet of floor area for each occupant of such unit in excess of two.
2. The unit must have a separate closet.
3. The unit must be provided with a kitchen sink, cooking appliance, and refrigerator, each having a clear working space of not less than 30 inches in front. Light and ventilation conforming to this code shall be provided.
4. A separate bathroom containing a water closet, lavatory, and bathtub or shower. ●

CHAPTER

6 Special Uses and Considerations

© International Code Council

Once a building has been classified to an occupancy group and type of construction, a more detailed design or review can be done. Buildings are not just a single occupancy and they are not just one- or two-story buildings. Many structures have multiple types of uses in them that may create a problem in handling the difference in hazards and relationships between the uses. With the cost of land increasing in our urban areas, more buildings are going vertical. High-rise buildings create a challenge for firefighting and egress, which are addressed in the codes. Innovative designs also create challenges for the safety of occupants in a building. Some buildings have a higher level of hazards that may affect people inside and outside the building, as well as the emergency responders. This chapter discusses some of the more common and interesting uses and design options permitted by the IBC.

FIGURE 6-1 Mixed occupancy building with Group M on the first floor and Group B on the second floor

MULTIUSE BUILDINGS

Most buildings are built with two or more uses, or occupancies, incorporated into the design. For example, shopping centers may have retail uses (Group M), business uses (Group B), restaurant uses (Group A-2), and other assembly uses (Group A-3) (see Figure 6-1). All of these uses have different hazards associated with them and may or may not be compatible with each other. The IBC addresses these differences by regulating how the building is constructed when multiple uses are included in a structure. The code outlines three options for dealing with multiple occupancies in a building. The three options are separated occupancies, nonseparated occupancies, and accessory occupancies. The code also allows the designer to combine these options in the same building. **[Ref. 508]**

Separated occupancies

When a building contains multiple uses, the designer can choose to separate the occupancies from each other by construction having a fire-resistance rating. When the designer chooses this option, the building is considered to be a *separated occupancy building*. The hourly ratings of the fire separation are based on the requirements of IBC Table 508.4. This table gives credit for the installation of a fire sprinkler system by reducing the required fire ratings of the separation by 1 hour. For example, if the shopping center in the preceding example was provided with fire sprinklers, it would be required to have a 1-hour fire-resistance rated separation between the two assembly occupancies (A-2 and A-3) and the other three occupancies (B, M, and S-1). However, a separation would not be required between the B, M, and S-1 occupancies

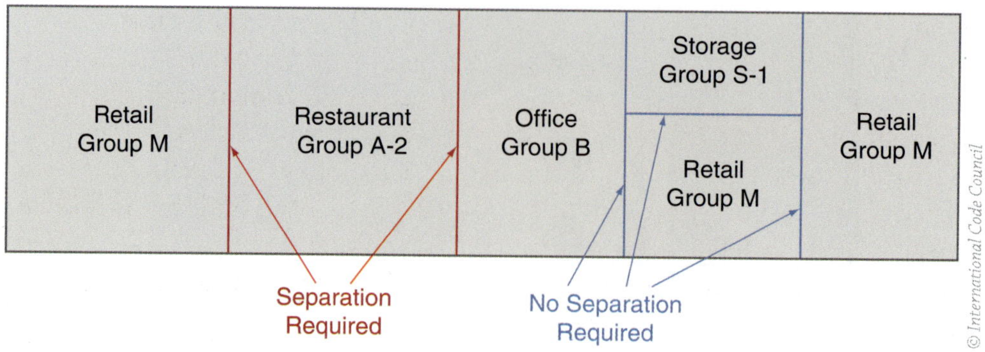

FIGURE 6-2 A Separated occupancy building

(Figure 6-2). When a separation is required in Table 508.4, it must be constructed as fire barriers (Section 707) and/or horizontal assemblies (Section 712).

When a building is designed as a separated occupancy, the allowable building area discussed in Chapter 5 must be expanded to address the multiple occupancies. The IBC requires that a calculation be done to confirm that the percentages of each of the different occupancies added together do not exceed 100 percent. A ratio calculation is done by dividing the actual area of each occupancy by the allowable area of each occupancy. All of the ratios or percentages are added together for each story and are not permitted to exceed 1, or 100 percent. This calculation is done for each individual story in the building. As long as none of the floor's ratios exceed 1 for a single-, two-, or three-story structure, the building is deemed to comply with the height and area requirements. To determine the maximum area of the building that is four or more stories in height, the ratios of all of the stories are added together. The total of all of these ratios cannot exceed 3. **[Ref. 508.4]**

Nonseparated occupancies

The second option is to classify the building as a *nonseparated occupancy building* (see Figure 6-3). This option is by far the most common approach to dealing with multiple occupancies in a building. When a

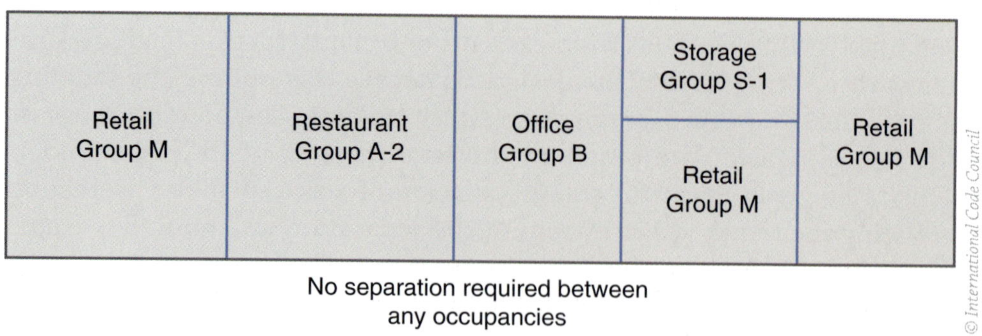

FIGURE 6-3 Nonseparated occupancy building

designer chooses this option, he or she must determine the most restrictive occupancy in a building based on height and area. The allowable area and height of each story are then based on that most restrictive occupancy. As long as the actual area of each story does not exceed the permitted area for the most restrictive occupancy in a one-, two-, or three-story building, the building complies. When the building is four or more stories in height, a ratio calculation similar to that for separated occupancies for each story must be done. The sum of all the story ratios must be 3 or less to show compliance.

With the exception of some Group H occupancies and residential-type occupancies, the different uses located in a nonseparated occupancy building are not required to be separated from each other by fire-resistance rated assemblies. However, when this option is selected, fire sprinklers and/or fire alarms may be required to be installed due to one of the occupancies. The IBC requires that the most restrictive requirements for fire sprinklers and alarms be applied to the entire building. This means that if any one of the uses in a nonseparated building is required to be provided with fire sprinklers or alarms, then the entire building must be protected by the required fire protection systems. In addition, each of the individual occupancies must comply with the specific requirements for each specific use. **[Ref. 508.3]**

Accessory occupancies

Accessory occupancies are those uses that are subsidiary to the main occupancy in a building. For example, a small office space in a factory could be considered an accessory occupancy as long as it complies with the provisions of Section 508.2. An accessory use must be given an occupancy classification. So, the office space in the example would be classified as a Group B occupancy. In order for a use to be considered an accessory occupancy, the space cannot have a floor area that is more than 10 percent of the total floor area on a story. There can be more than one accessory use, or more than one accessory use area, on a story. The aggregate area of all the accessory uses cannot exceed 10 percent of the story. The area is also limited to the floor area and height limits in Table 503 without any increases. The allowable area for the building is calculated based on the occupancy classification of the main occupancy. In the example, the factory would be classified as a Group F-1 occupancy. If the design complies with these requirements, the accessory occupancy is not required to be separated by a fire-resistance-rated wall or floor system from the main occupancy (see Figure 6-4). **[Ref. 508.2]**

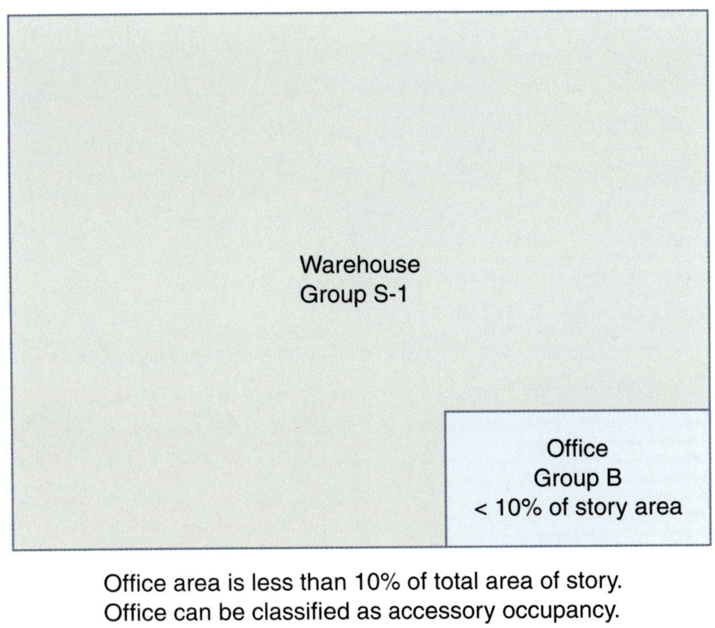

FIGURE 6-4 Example of accessory occupancy area

Incidental Uses

Another area in a building that needs to be addressed is an incidental use. These areas are ancillary functions that generally pose a greater level of hazard in a particular occupancy. These areas require a higher level of protection to protect the occupants in the main occupancy from the hazards that may exist in an incidental area. The IBC requires that incidental uses be separated from the rest of the building by either a fire-resistance-rated separation or be provided with automatic fire sprinklers (Table 6-1). If the separation option is selected, the separation must be either a fire barrier in accordance with IBC Section 707, or horizontal assemblies in accordance with IBC Section 712, or both.

When the fire sprinkler option is used, the room must be constructed to resist the passage of smoke. The walls must extend from the floor to either a fire-resistance rated horizontal assembly or the floor or roof sheathing above. Doors must be self- or automatic-closing and cannot have any air transfer openings in them. In addition, any air transfer openings in the enclosure wall must be provided with a smoke damper (Figure 6-5). **[Ref. 508.2.5]**

HIGH-RISE BUILDINGS

High-rise buildings are defined as having an occupied floor more than 75 feet above the lowest level of fire department vehicle access. The

TABLE 6-1 Incidental uses (Based on IBC Table 509)

Room or Area	Separation and/or Protection
Furnace room where any piece of equipment is over 400,000 Btu per hour input	1 hour or provide automatic fire-extinguishing system
Rooms with boilers where the largest piece of equipment is over 15 psi and 10 horsepower	1 hour or provide automatic fire-extinguishing system
Refrigerant machinery rooms	1 hour or provide automatic sprinkler system
Hydrogen cut-off rooms, not classified as Group H	1 hour in Group B, F, M, S, and U occupancies. 2 hours in Group A, E, I, and R occupancies.
Incinerator rooms	2 hours and automatic sprinkler system
Paint shops, not classified as Group H, located in occupancies other than Group F	2 hours; or 1 hour and provide automatic fire-extinguishing system
Laboratories and vocational shops, not classified as Group H, located in Group E or I-2 occupancies	1 hour or provide automatic fire-extinguishing system
Laundry rooms over 100 square feet	1 hour or provide automatic fire-extinguishing system
Group I-3 cells equipped with padded surfaces	1 hour
Group I-2 waste and linen collection rooms	1 hour
Waste and linen collection rooms over 100 square feet	1 hour or provide automatic fire-extinguishing system
Stationary lead-acid battery systems having a liquid capacity of more than 100 gallons used for facility standby power, emergency power, or uninterrupted power supplies	1 hour in Group B, F, M, S, and U occupancies. 2 hours in Group A, E, I, and R occupancies.

FIGURE 6-5 Room with a boiler greater than 15 psi is an incidental use

lowest level of fire department vehicle access is the lowest elevation that a fire truck can park its apparatus adjacent to the building and then extend the ladder to access the upper floors (Figure 6-6). **[Ref. 202]**

High-rise buildings introduce serious challenges for occupants and firefighters in emergency situations. Manual firefighting equipment cannot reach the higher floors. Therefore, the buildings must be provided with fire sprinklers and fire standpipes. Standpipes extend the water supply that firefighters need to fight the fire from the connection to the public water supply to valves located within the building's exit stairs.

High-rise buildings are also provided with other emergency systems. Smoke detectors are required to be installed in elevator lobbies and elevator machine rooms, as well as in mechanical, electrical, and similar

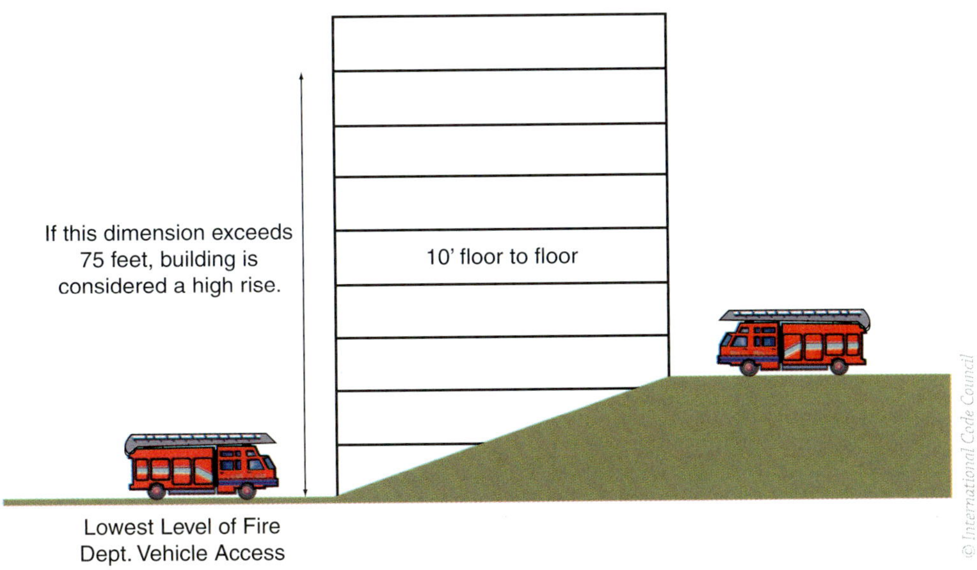

FIGURE 6-6 Determination of high-rise building

rooms when they are not protected by the automatic sprinkler system. An emergency voice/alarm system must be provided to alert the occupants of an emergency situation and give directions on how to evacuate the building. Standby and emergency power systems are also required to be installed in high-rise buildings. These systems maintain power for emergency lighting, ventilation, and elevators. They also maintain electrical power to the fire alarm and communication systems, as well as to electrically powered fire pumps. **[Ref. 403.4]**

Getting people out of a high-rise building during an emergency situation is difficult to address. As people are trying to egress the building, firefighters are trying to access the floor or floors where the fire is burning. Therefore, the IBC requires that the stairs serving as the means of egress be enclosed in a "smokeproof enclosure." These enclosures are constructed to prevent smoke from getting into the stair enclosure when the doors are opened by the occupants or firefighters. In buildings over 420 feet in height, an additional stairway must be provided over and above the required number of stairways required by Chapter 10 of the IBC. The purpose of these extra stairways is to provide firefighters with an additional stair to access the floors that they need to get to. The

FIGURE 6-7 High-rise building under construction

IBC does, however, allow the substitution of an elevator used for occupant self-evacuation for the additional stairway. **[Ref. 403.5]**

To facilitate the rapid deployment of firefighters, high-rise buildings with an occupied floor over 120 feet above the lowest level of fire department access require a minimum of two fire service access elevators. These elevators are located directly adjacent to an exit enclosure and can be used for the deployment of firefighters to floors near the fire's area of origin and for rescue of individuals within areas of refuge or who need medical treatment. Special features for this elevator are specified in Section 3007. **[Ref. 403.6]**

When fires do occur in a high-rise building, smoke accumulates in a building and is difficult to remove. Therefore, the IBC requires a smoke removal system. The purpose of these systems is to facilitate the smoke removal in post-fire salvage and overhaul operations. The designer has two basic options for complying with this requirement. The first option is to install manually operable windows around the perimeter of the building. The second option is a mechanical air-handling system that can exhaust an area directly to the outside within fifteen minutes. The designer has one other option to engineer a system to produce equivalent results. **[Ref. 403.4.6]**

To provide an additional level of safety in exit stairways, the IBC requires luminous egress path markings that delineate the exit path in high-rise buildings that contain Groups A, B, E, I, M, and R-1 occupancies. These markings are glow-in-the-dark materials that are intended to provide occupants with increased visibility in stair enclosures and exit passageways. They are required on the leading edge of stairs, landings, and handrails; around the perimeter of floors and wall, including any transitions; obstacles at or below 6 feet, 6 inches; and doors leaving the exit enclosure (Figure 6-8). **[Ref. 1024]**

© International Code Council

FIGURE 6-8 Self-luminous markings in high-rise building stairway

ATRIUMS

Requirements for atriums were introduced into the codes in the early 1980s. Building designers wanted to have floors open to each other over several stories. An *atrium* is typically an opening between three or more floors of a building, although it can be open to two floors in some instances. Atriums create challenges similar to high rises in that it is difficult for firefighters to control fires within them. Atriums are essentially chimneys within the building that people pass through, sometimes during an emergency situation. Therefore, smoke control is an important aspect of an atrium design (Figure 6-9). **[Ref. 404]**

The first part of controlling smoke is to limit the amount of smoke developed in a fire. Therefore, buildings that contain an atrium are required to be provided with fire sprinklers throughout. By suppressing the fire with sprinklers, the amount of smoke is limited. The next step is to provide a smoke-control system that provides a safe path of egress for the occupants to egress the building. Unlike the smoke-removal provisions in a high-rise building, this system is not intended to be used for post-fire operations. A typical atrium smoke-control system creates a

FIGURE 6-9 Three-story atrium

negative pressure within the atrium and exhausts the smoke out through openings, providing a safer atmosphere for people to egress. **[Ref. 404.5 and 909]**

Areas of the building that are not included in the design of the smoke-control system must be separated from the atrium with a minimum of 1-hour fire-resistance rated walls called *fire barriers*. These assemblies are discussed later. This separation also assists in controlling the smoke in a fire situation. To maintain the operation of the smoke-control service, a standby power system is required to be provided when an atrium is included in a building design. **[Ref. 404.6]**

VEHICLES IN BUILDINGS

Anytime you place a motor vehicle inside a building, you increase the potential hazards in the building. Inside buildings, vapors from gasoline or other fuels can accumulate and create a fire or explosion hazards. Vehicle engine exhaust from operating vehicles can displace oxygen in the air and create serious health hazards for the building occupants. The IBC addresses these hazards by regulating the design and construction of parking garages and repair garages.

Private garages

Chapter 3 discussed the classification of private garages and carports as Group U, Utility Occupancies. These uses are intended to include single- or double-car garages that serve residential dwelling units. They are limited in size to 3,000 square feet when private or pleasure-type vehicles are stored in the building. The code also presumes that no repair work will be conducted on the vehicle in the garage.

The separation between private garages and other dwelling units is less restrictive than other occupancy separations. The separation between the garage and dwelling unit is only required to have ½-inch-thick gypsum board applied on the garage side of the separation. However, if the separation is a floor-ceiling assembly located below habitable space, then ⅝-inch-thick Type X gypsum board is required to be applied to the ceiling. Openings in the separation are also limited. Doors must be either 1⅜-inch-thick solid wood, honeycomb steel, or 20-minute fire-rated. The door must be installed so that it is self-closing. **[Ref. 406.3]**

Parking garages

Parking garages are classified as either enclosed or open (Figure 6-10). Enclosed parking garages have a higher chance of accumulation of vapors and exhaust gases. Therefore, they must be equipped with a mechanical ventilation system that maintains safe air quality levels. Enclosed parking garages are also required to be provided with a fire sprinkler system to control any fires that may occur within the vehicles. The height of an area of an enclosed parking garage is regulated by Section 503, like other buildings. **[Ref. 406.4]**

Open parking garages are provided with natural ventilation by means of openings in the exterior walls. The IBC requires that a minimum of

FIGURE 6-10 Open parking garage

FIGURE 6-11 Repair garage

20 percent of the total perimeter wall area of each parking level be open on at least two sides of the garage. These openings allow the natural movement of air to ventilate the fumes and gases that accumulate in the garage. Open parking garages are required to be constructed of Type I, II, or IV construction. If an open parking garage is attached to another use, the height and area are regulated by Section 503. However, if it is a separate building used exclusively for the parking or storage of private motor vehicles, it is permitted to be larger in size. **[Ref. 406.4]**

Repair garages

A *repair garage* is defined in the *International Fire Code* (IFC) as "a building, structure, or portion thereof used for servicing or repairing motor vehicles." This can include a stereo installation bay, a transmission shop, or an auto-body shop (Figure 6-11). These uses may result in the release of flammable vapors, but occupants also can be exposed to some of the materials and processes that are used in the repair of vehicles. Solvents, oils, and other hazardous materials must be addressed in these occupancies. The amounts of these materials are limited within Group S-1, Repair Garages. If the operator exceeds those limits, the use must be classified as a Group H occupancy. Repair garages are required to be provided with mechanical ventilation similar to parking garages. **[Ref. 406.8]**

As technology improves in the design and production of automobiles, fuels other than gasoline are being used. Many new designs are using nonodorized gases like hydrogen or liquefied natural gas (LNG). These gases are lighter than air. Therefore, the conventional exhaust systems designed to remove gasoline or diesel vapors will not work. Therefore, special ventilation systems must be provided in repair garages utilizing these fuels. The system is activated by a gas detection system when the level of flammable gas exceeds 25 percent of the lower explosive limit. **[Ref. 406.8.5]**

Hospitals

Hospitals are unique uses in that many people are in the building, and they may be nonambulatory, unconscious, or otherwise incapacitated.

Therefore, the code provides specific requirements to address the unique conditions that exist in a hospital. Hospitals are divided into smoke compartments by smoke barriers. The smoke compartments are limited to 22,500 square feet in area. Smoke barriers are discussed in Chapter 8. The purpose of the smoke compartment is to provide what is called "protection in place." This means that if a fire occurs in one smoke compartment, patients can be relocated into an adjacent smoke compartment and protected from the fire. Hospital personnel are trained to respond to emergencies in this manner. Each smoke compartment is designed to include areas that are sized to accommodate the patients from the adjacent smoke compartment. **[Ref. 407.4]**

HAZARDOUS MATERIALS

The IBC regulates buildings that contain hazardous materials that are either a physical hazard or health hazard. Physical hazard materials are materials that burn, have accelerated burning characteristics, and/or produce an explosion. Health hazard materials can cause death or injury or can incapacitate a person after a single, brief exposure.

When these materials are manufactured, stored, or used in a building, the occupants must be protected from the potential hazards. Any time a building contains hazardous materials, a report must be provided to the building official for review. It must be prepared by a qualified person, firm, or corporation approved by the building official. This report must classify the hazardous materials using the criteria in the IBC and IFC and must indicate the maximum amounts of hazardous materials that will be stored and/or used in the building. It must also explain the type and level of protection that is necessary to safely store and handle these materials inside the building.

Maximum allowable quantities

The IBC limits the amount of hazardous materials in uses other than Group H occupancies. Once the limits in Section 307.1 are exceeded, the building can be classified as a Hazardous Group H occupancy. The quantity limits are based on how the hazardous materials are either stored or used in a building. If the hazardous material is used in such a way that it is continuously open to the atmosphere during normal operations and vapors are released, it is considered to be an open system. When the material is kept in a closed vessel or system and vapors are not released, it is called a *closed system*. The IBC regulates the maximum hazardous material quantities based on the environment.

The maximum amounts of materials are outlined in Table 307.1(1) and 307.1(2). A sample of this table is shown in Table 6-2. This limit is the maximum allowable quantity (MAQ) per control area. The MAQ is based on the hazard classification of the material, its situation (storage or use), and the material's physical state. There are several footnotes to the table, but the most common ones are footnotes d and e. These two footnotes allow the designer to increase the MAQs when the building

TABLE 6-2 Maximum allowable quantity per control area of hazardous materials posing a physical hazard (Based on a portion of IBC Table 307.1(1))

Material	Class	Storage		
		Cubic Feet	Liquid Gallons (pounds)	Gas (cubic feet @ NTP)
Combustible liquid	II IIIA IIIB	N/A	120 330 13,200	N/A
Combustible fiber	Loose Baled	100 1,000	N/A	N/A
Flammable gas	Gaseous Liquefied	N/A	N/A (150)	1,000 N/A

is provided with fire sprinklers and/or the materials are stored in an approved manner outlined in the footnotes. In most cases, the MAQs are permitted to be increased 100 percent in each case and can be accumulated with each other. **[Ref. 307.1]**

Code Basics

Determining the MAQ in One Control Area

Given: A one-story building will be used for the storage of toluene, a Class IC flammable liquid, a physical hazard material. The building is protected throughout by an automatic fire sprinkler system complying with Chapter 9 and is one control area.

Problem: Determine the MAQ using the requirements in Table 307.1(1).

Solution:

MAQ from Table 6-3 for Class IC, flammable liquid = 120 gallons

Increase for installation of fire sprinklers (footnoted) = 100 percent

120 × 2 = 240 gallons permitted in control area ●

Control areas

One way to increase the amount of hazardous materials within a building is to place them in spaces that are separated from the rest of the building by fire-resistance rated fire assemblies. Under the IBC and IFC hazardous material provisions, all buildings are considered a single control area. A building can have two or more control areas based on the location of the storage and use in relation to the building's height. The maximum numbers of control areas and separation requirements are shown in Table 6-3. The percentage of the maximum allowable quantity (MAQ) of materials is limited based on the height of the building. **[Ref. 414.2]**

If the amount of hazardous materials exceeds the MAQ within a control area, the building is classified as a Group H occupancy. If the amounts are within the MAQ, the building is classified as an occupancy that it most resembles. For example, if a repair garage stores a Class II combustible liquid in the building and the total amount that they store exceed the MAQ, it would be classified as a Group H-2 or H-3. If it does not exceed the MAQ, the building would be classified as a Group S-1 occupancy. However, the storage of the material would still need to comply with the requirements of Section 414 regarding ventilation and emergency alarms.

When buildings are classified as Group H occupancies, additional requirements are applied to the design based on the physical and health hazards of the stored and used materials. This may include spill control and containment of liquid hazardous materials. It may also require specific separation of the Group H occupancy from other occupancies. In the case of a Group H-1 occupancy storing explosives in quantities in excess of the MAQ, the building must be a separate building that cannot be attached to other buildings. The buildings must be placed a specific distance from other buildings based on the hazard classification and quantity of explosive materials. **[Ref. 415.3]**

TABLE 6-3 Design and number of control areas (Based on IBC Table 414.2.2)

Floor Level		Percentage of the Maximum Allowable Quantity per Control Area[a]	Number of Control Areas per Floor	Fire-resistance Rating for Fire Barriers in Hours[b]
Above grade plane	Higher than 9	5	1	2
	7–9	5	2	2
	6	12.5	2	2
	5	12.5	2	2
	4	12.5	2	2
	3	50	2	1
	2	75	3	1
	1	100	4	1
Below grade plane	1	75	3	1
	2	50	2	1
	Lower than 2	Not Allowed	Not Allowed	Not Allowed

a. Percentages shall be of the maximum allowable quantity per control area shown in Tables 307.1(1) and 307.1(2), with all increases allowed in the notes to those tables.

b. Fire barriers shall include walls and floors as necessary to provided separation from other portions of the building.

LIVE/WORK UNITS

A new type of development is being constructed in many parts of the country in which a person can buy or lease a commercial space with a living unit above or adjacent to it. These spaces are defined as *live/work units* in the IBC (Figure 6-12). Live/work units are limited in many ways due to the potential hazards that may exist in the nonresidential portion. These hazards can significantly impact the adjacent residential units. They are limited to a maximum area of 3,000 square feet; the non-residential area cannot exceed 50 percent of the unit; the nonresidential portion of the

FIGURE 6-12 Live/work units

unit must be located on the first or main floor; and they are limited to a maximum of five nonresidential workers or employees. [Ref. 419]

Live/work units are classified as Group R-2 occupancies. A separation is not required between the residential and nonresidential portion of the unit because the entire unit is classified as a Group R-2 occupancy. In addition, some of the exiting requirements for residential use are permitted of the nonresidential use. The units may need to be provided with a monitored fire alarm system and must be provided with an automatic fire sprinkler system. They must also meet the accessibility requirements of the IBC.

AMBULATORY HEALTH-CARE FACILITIES

Ambulatory health-care facilities are buildings or spaces that provide medical, surgical, psychiatric, nursing, or similar care on less than a 24-hour basis to individuals who are not capable of saving themselves during an emergency (Figure 6-13). An outpatient surgery center is an example of an ambulatory health-care facility. When a person is receiving this type of care, he or she may be under a general anesthesia and not be able to get up and out of the building. Therefore, additional safeguards must be provided to protect the patients in this condition.

Ambulatory health-care facilities are classified as Group B occupancies. If a facility has the potential of having four or more patients that are not capable of saving themselves, it must be separated from other adjacent spaces, corridors, or tenants with fire partitions, as described in Chapter 8. Facilities that exceed 10,000 square feet must be subdivided into two smoke-protected compartments. The separation is provided by a smoke barrier wall constructed in accordance with IBC Section 710. Areas on each side of the smoke barrier must be provided to act as refuge areas for patients if a fire should occur. The intent of this requirement is to protect the patients in place by moving them into a separate smoke compartment away from the fire. In addition, the ambulatory health-care facility must be provided with an automatic sprinkler system, even though Group B occupancies are not typically required be provided with such a system. If the automatic sprinkler system is only installed within the facility, the public use areas of the rest of the building must be provided with an electronically supervised automatic smoke-detection system. [Ref. 422]

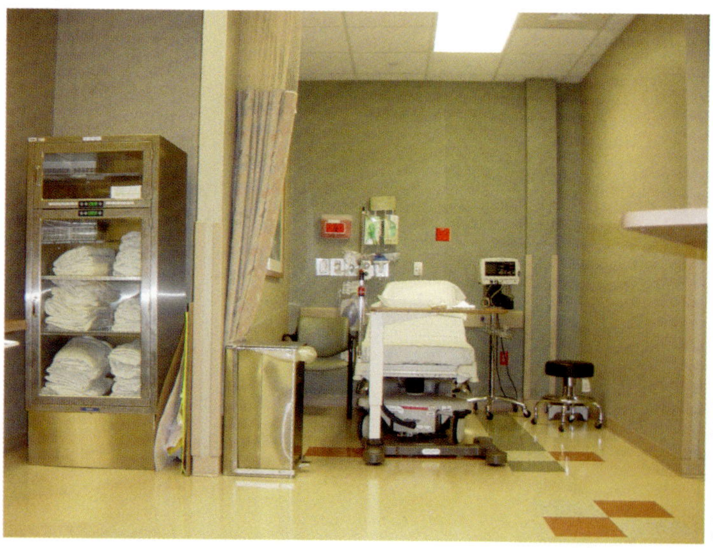

FIGURE 6-13 Recovery room of an ambulatory health-care facility

PART III

Fire Safety

CHAPTER

7 Fire and Smoke Protection

One of the leading causes of death in buildings is attributed to fire and smoke. Fire can spread rapidly within a building and, in some cases, from building to building. The *International Building Code* is designed to limit the spread of fire inside and outside of a building. The IBC protects the structure and occupants from fire by providing both passive and active fire protection. *Passive fire protection* safeguards the building and occupants from fire by constructing the fire protection into the structure. This is done by protecting the structural frame of the building and constructing fire-resistance rated construction walls, floors, and roofs. *Active fire protection* is commonly provided by automatic fire sprinklers. These systems provide protection to the building and/or occupants by suppressing a fire that may occur in a building. This part discusses the way the building code provides protection from fire spreading within a building and from building to building. It then describes the different ways that a building is protected from fire.

FIGURE 7-1 Fire damage to a four-story apartment building

SPREAD OF FIRE WITHIN A BUILDING

The spread of fire within a building is a serious issue. A number of different hazards can exist in a structure. There may be different occupancies in buildings that need to be separated from each other. Many buildings have rooms or areas that are not considered different occupancies, but contain uses or equipment that pose a hazard to the occupants. The condition or number of occupants in a building can also have a significant impact on fire protection. The building may need to be divided into compartments to protect the people from a fire event.

The IBC provides protection from fire spread in a building by requiring specific fire-resistance-rated separations between hazards that may be located in the building. It also protects people as they try to evacuate a building in a fire situation. In addition to protecting occupants from fire, the IBC also provides protection for firefighters entering a building during a fire event (Figure 7-1).

The IBC requires fire-resistant separations within buildings and protection of the structure with fire-resistance-rated protection. The separations differ in the level of protection and continuity. Fire-resistance-rated assemblies can be both vertical and/or horizontal. Openings in a fire-resistant-rated assembly can allow fire and smoke to spread from area to area. Therefore, doors and windows located in an assembly are required to be provided with a fire protection rating as well. Pipes, cables, and ducts penetrating the fire assemblies that are not sealed properly can also significantly affect the fire-resistance rating of an assembly. So, these penetrations must be sealed with an approved material that resists the fire and smoke from passing through these small gaps in the assembly.

FIGURE 7-2 1871 Great Chicago fire

SPREAD OF FIRE FROM BUILDING TO BUILDING

The Great Chicago Fire in 1871 revealed that we needed to protect buildings from spread of fire from neighboring buildings. The spread of fire outside of the structure can have devastating effects on a community (Figure 7-2). The IBC regulates the spread of fire by increasing the fire resistance of exterior walls as they are placed closer to a lot line. Windows, doors, and ducts located in those exterior walls are also limited the closer the building gets to the lot line. This protection contains the fire on the property on which it originates and limits the impact on adjacent buildings. It can also protect a building from a fire in an adjacent building.

The IBC addresses the spread of fire from building to building by regulating the construction of exterior walls in relationship to the lot line. The closer the distance the exterior wall is to the lot line, the more restrictive the fire-resistance rating must be to protect the adjacent property from the radiant heat generated during a fire. Openings and penetrations must also be protected in exterior walls when located specific distances from the lot line. Projections extending from the exterior wall are also restricted based on their location's relation to the lot line.

CHAPTER

8 Passive Fire Protection

© International Code Council

As noted in the previous chapter, passive fire protection is designed to provide safeguards from fire by building the protection into the building construction. This chapter discusses the different types of passive fire protection that the *International Building Code* specifies. Each of the different fire assemblies has a specific purpose. It is important that the user of the code be familiar with the different terms used in the IBC. This chapter also provides the way that fire-resistant assemblies are determined through standard testing and documentation.

DETERMINATION OF FIRE-RESISTANCE-RATED CONSTRUCTION

There are several ways to determine the fire-resistance rating of different assemblies. Fire-resistance-rated assemblies include floor/ceiling assemblies, roof/ceiling assemblies, wall assemblies, door assemblies, and window assemblies. The most common way to determine the fire-resistance rating is by subjecting the assembly to a specific test. Section 721 of the IBC provides several prescriptive assemblies that have proven themselves over time and with early testing. The ratings can also be calculated using known factors of historical materials.

Testing

Fire-resistance-rated construction is typically designated by the number of hour(s) that an assembly will resist fire on one side. The hourly rating is based on a standardized test developed by the American Society for Testing and Materials (ASTM) and Underwriters Laboratories™ (UL). ASTM Standard E119 and UL263 are the standards used to test structural members, walls, and horizontal assemblies (floor/ceiling and roof/ceiling assemblies). These standards ensure that fire-resistant properties of materials and assemblies are measured and specified to a common standard. They prescribe a standard exposing fire of controlled extent and severity to the assembly. Therefore, all types of materials and assemblies are tested to the same standard. The test is conducted by an approved third-party testing agency. **[Ref. 703.2]**

The standards use a time–temperature curve to evaluate the fire-resistance rating of an assembly (Figure 8-1). To get a fire rating, the manufacturer of an assembly constructs it just like it will be built in a building. The testing agency then places the assembly in a test furnace with gas jets that simulate a fire on one side of an assembly. If an assembly is designed to support a load, the test includes the application of such load on the load-bearing member, wall, or horizontal assembly.

Thermocouples are placed on the opposite sides of walls and horizontal assemblies to test the temperature on the unexposed side of the assembly. The standard sets specific temperatures that the thermocouples register for different types of assemblies. In addition, some wall assemblies must undergo what is called a *hose-stream test*. In this test, a duplicate assembly is exposed to the fire-endurance test and then sprayed with a fire hose at a specific pressure and duration. As long as the hose stream does not create openings in the assembly beyond the unexposed side of the assembly, the test is accepted (Figure 8-2).

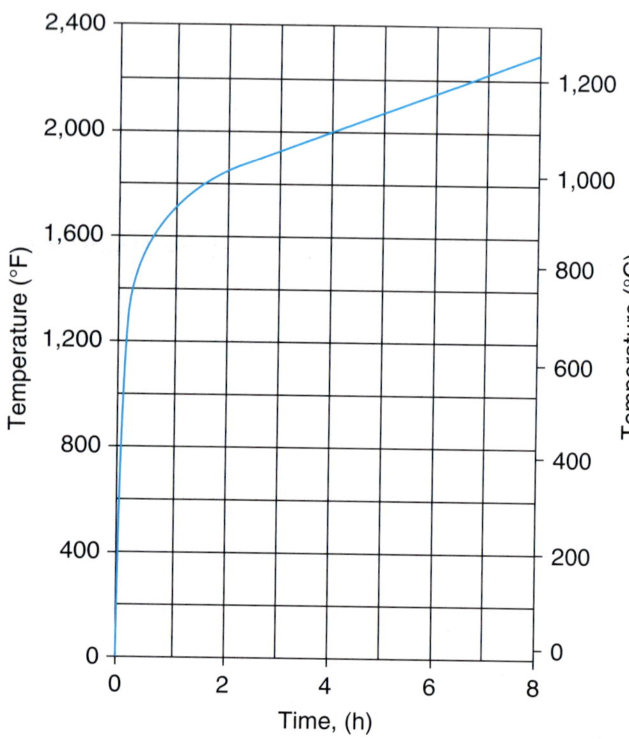

FIGURE 8-1 ASTM E119 time–temperature curve

FIGURE 8-2 Floor/ceiling test furnace at testing facility

TABLE 8-1 Minimum protection of structural parts based on time periods for various noncombustible insulating materials (A portion of IBC Table 721.1(1))

Structural Parts to be Protected	Item Number	Insulating Material Used	Minimum Thickness of Insulating Material for the Following Fire-Resistance Periods (inches)			
			4 Hours	3 Hours	2 Hours	1 Hour
5. Reinforcing steel in reinforced concrete columns, beams, girders, and trusses	5–1.1	Carbonate, lightweight, and sand-lightweight aggregate concrete, members 12^2 or larger, square, or round. (Size limit does not apply to beams and girders monolithic with floors.)	1½	1½	1½	1½
		Siliceous aggregate concrete, members 12^2 or larger, square or round. (Size limit does not apply to beams and girders monolithic with floors.)	1½	1½	1½	1½
6. Reinforcing steel in reinforced concrete joists	6–1.1	Carbonate, lightweight, and sand-lightweight aggregate concrete.	1	1	¾	¾
	6–1.2	Siliceous aggregate concrete.	1¼	1	1	¾

Documentation

Once an assembly passes the ASTM E119 or UL 263 test, the results are published by the testing agency. In some cases, the test has been conducted so many times and with similar materials that the assemblies are listed in the IBC. The code provides prescriptive descriptions of fire-resistance-rated components and assemblies. Fire protection systems are listed in Table 721.1 (1), (2), and (3) for components and assemblies that have historically been accepted as fire-resistance-rated construction. Examples of these assemblies are shown in Table 8-1 through Table 8-3. These tables are excerpts of the tables in Section 721.1.

For assemblies not included as prescriptive systems in the IBC, the testing agencies publish their test results for building departments to use.

TABLE 8-2 Rated fire-resistance periods for various walls and partitions (Based on IBC Table 721.1(2))

Material	Item Number	Construction	Minimum Finished Thickness Face-to-Face (inches)			
			4 Hours	3 Hours	2 Hours	1 Hour
13. Noncombustible studs—interior partition with gypsum wallboard each side	13–1.1	0.018^2 (No. 25 carbon sheet steel gage) channel-shaped studs 24^2 on center with one full-length layer of 5/8^2 Type X gypsum wallboard applied vertically attached with 1^2 long No. 6 drywall screws to each stud. Screws are 8^2 on center around the perimeter and 12^2 on center on the intermediate stud. The wallboard may be applied horizontally when attached to 35/8^2 studs and the horizontal joints are staggered with those on the opposite side. Screws for the horizontal application shall be 8^2 on center at vertical edges and 12^2 on center at intermediate studs.	–	–	–	2⅞
14. Wood studs—interior partition with gypsum wallboard each side	14–1.3	2^2' 4^2 wood studs 24^2 on center with 5/8^2 Type X gypsum wallboard applied vertically or horizontally nailed with 6d cooler or wallboard nails at 7^2 on center with end joints on nailing members. Stagger joints each side.	–	–	–	4¾

TABLE 8-3 Minimum protection for floor and roof systems (A portion of IBC Table 721.1(3))

Floor or Roof Construction	Item Number	Ceiling Construction	Thickness of Floor or Roof Slab (inches)			
			4 Hours	3 Hours	2 Hours	1 Hour
1. Siliceous aggregate concrete	1–1.1	Slab (no ceiling required). Minimum cover over nonprestressed reinforcement shall not be less than ¾"	7.0	6.2	5.0	3.5
2. Carbonate aggregate concrete	2–1.1		6.6	5.7	4.6	3.2
3. Sand-lightweight concrete	3–1.1		5.4	4.6	3.8	2.7
4. Lightweight concrete	4–1.1		5.1	4.4	3.6	2.5

The more common testing agencies used include Underwriters Laboratories™ (UL), Intertek ETL Semco, and Factory Mutual (FM). When assemblies listed by these agencies are used in a building, they must be constructed in strict accordance with the agency's listing. If they are not constructed in this manner, the assembly may not perform in a fire as expected.

TYPES OF FIRE ASSEMBLIES

There are several different fire assemblies and components in a building. There are vertical assemblies, horizontal assemblies, and structural frame members. The requirements are different for each type of assembly. It is important that the code user use the proper terminology when discussing the code. If the incorrect terms are used, improper design or enforcement can occur. This chapter discusses the major fire assemblies that can be found in a building.

Protection of the structure

The stability of the building is important to protect the occupants trying to evacuate as well as the firefighters entering the building to extinguish a fire. Therefore, the IBC regulates how the structural elements of a building are to be protected from fire. As previously discussed, Chapter 6 of the IBC discusses the types of construction of a building and outlines the fire protection of the building elements that are essential to the stability of a building. Table 601 of the IBC outlines those protection requirements.

Structural frame members are separated into two different categories, primary and secondary members. Primary structural frame members are columns; beams, girders, and trusses directly attached to columns; floor and roof construction connected directly to columns; and bracing members that are designed to support the gravity loads in a building (see Figure 8-3). Secondary members are any structural members not connected directly to columns; floor and roof construction not connected to columns; and bracing members not supporting gravity loads (see Figure 8-4). **[Ref. 202]**

FIGURE 8-3 Column and girder truss are considered primary members

Structural frame members constructed of concrete and masonry provide the required fire protection by insulating the steel reinforcement with concrete or masonry. The thickness of the protection is listed in Table 721.1 as previously noted. Steel structural members are typically protected by spraying a fire-resistant material to insulate the steel from the heat of the fire. The material must be installed in accordance with the manufacturer's installation instructions and the material's approved listing (see Figure 8-5).

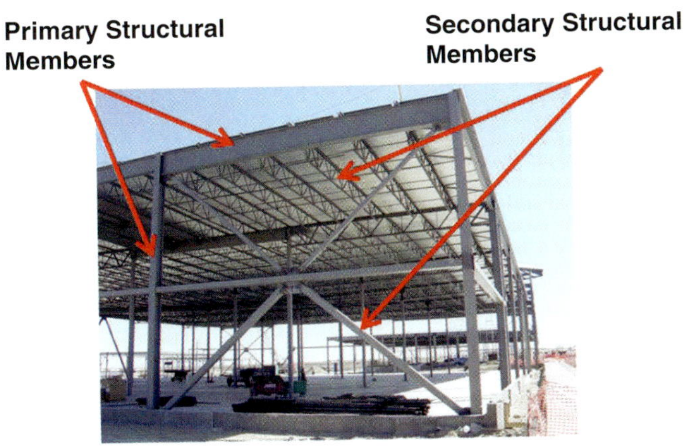

FIGURE 8-4 Primary and secondary structural members

FIGURE 8-5 Sprayed applied fireproofing on structural members

Exterior walls

We learned from the 1871 Chicago fire that it is important to limit the spread of fire from building to building. Section 602 specifies the minimum fire rating of exterior walls based on the *fire separation distance* (see Table 8-4). Fire separation distance is the dimension between the face of the exterior wall to an interior lot line; the center of a street, alley, or public way; or an imaginary line between two buildings on the same property (see Figures 8-6 and 8-7). When two buildings are located on the

TABLE 8-4 Fire-resistance-rating requirements for exterior walls based on fire separation distance (Based on IBC Table 602)

Fire Separation Distance = X (feet)	Type of Construction	Occupancy Group H	Occupancy Group F-1, M, S-1	Occupancy Group A, B, E, F-2, I, R, S-2, U
X < 5	All	3	2	1
5 ≤ X < 10	IA	3	2	1
	Others	2	1	1
10 ≤ X < 30	IA, IB	2	2	1
	IIB, VB	1	1	0
	Others	1	1	1
X ≥ 30	All	0	0	0

FIGURE 8-6 Exterior wall (right) located on lot line

same lot, the imaginary line can be placed anywhere between the two buildings. It does not need to be placed down the center. If, however, the combined areas of the two buildings are less than the allowable height and areas permitted in Chapter 5 of the IBC, the fire separation is no longer required. [Ref. 705.2]

Exterior walls are also required to be fire-resistance rated based on the type of construction. Table 601 outlines the requirements for exterior bearing walls. The rating of exterior walls is based on the more restrictive requirements between the type of construction and the fire separation distance requirements.

Openings in exterior walls allow fire to radiate from one building to another. In some cases, they permit fire to enter a structure from an adjacent building. The IBC limits the amount of fire spread in this way by restricting the amount and type of openings in an exterior wall. Openings include doors, windows, air intakes, scuppers, or vents. Table 705.8 outlines the percentage of openings permitted in exterior walls. The openings are divided between protected and unprotected openings. Protected openings are those with a fire-resistance rating in accordance with Table 716.5 for fire doors and shutters, and Table 716.6 for fire window assemblies. [Ref. 705.8]

Buildings often have eaves, cornices, balconies, and other architectural appendages extending out from the exterior walls. These appendages are considered projections in the IBC. Projections keep the heat from a fire from escaping from the building and upward. Therefore, the code restricts the extension of the projection from the building in relation to the fire separation distance (see Table 8-5). This prevents the buildup of fire and smoke that may affect the adjacent building during a fire. In some cases, the projection must be protected by 1-hour fire-resistant-rated construction, heavy timber construction, fire-retardant-treated wood, or as required by Section 1406.3. [Ref. 705.2]

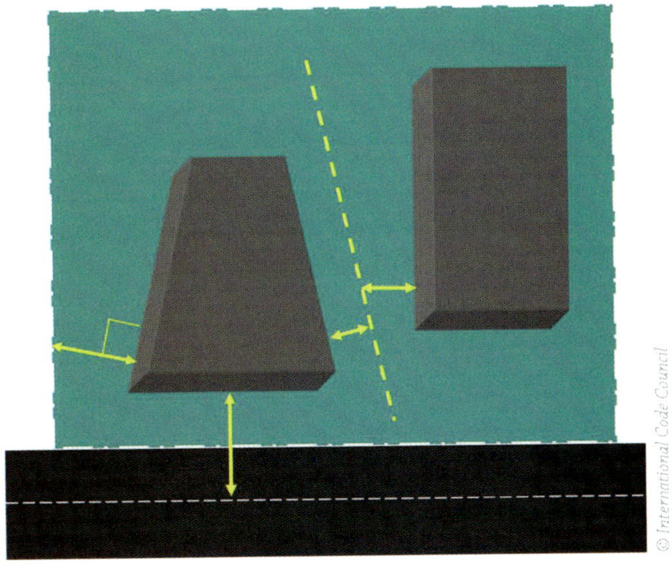

FIGURE 8-7 Fire separation distance measurement

Fire walls

Chapter 5 of the IBC limits the size of buildings based on the occupancy classification and type of construction as discussed earlier. One option available to the designer to increase the size of a structure is to construct a "fire wall." This may allow the building to be constructed to a lower type

Code Basics

Given:
Group R-2 occupancy
Type VA construction
35-foot fire separation distance

Problem:
Determine exterior bearing wall fire-resistance rating.

Solution:
Table 601 requires 1-hour construction.
Table 602 permits exterior walls to be nonrated.
Exterior wall must be 1-hour fire-resistance-rated construction based on type of construction.

TABLE 8-5 Minimum distance of projection (Based on IBC Table 705.2)

Fire Separation Distance (FSD)	Minimum Distance from Line Used to Determine FSD
0 feet to less than 2 feet	Projections not permitted
2 feet to less than 5 feet	24 inches
5 feet or greater	40 inches

FIGURE 8-8 A fire wall creates two separated buildings

of construction. When a fire wall is constructed in accordance with the IBC, each side of the structure on each side of the wall is considered a separate building (see Figure 8-8). Therefore, the size of the building can theoretically be doubled and the type of construction maintained. A fire wall can also serve as a separation of occupancies or to separate different types of construction. A fire wall is the most important fire assembly in a building. **[Ref. 706.1]**

Fire walls have specific construction requirements. The most difficult provision is the requirement for structural stability. The wall must be constructed so the buildings on each side of the wall can collapse under fire conditions in such a way that the wall is not affected. This can be done in many different ways and is left up to the designer to show how this performance-based requirement is being met. Fire walls must extend from the foundation to a point 30 inches above both adjacent roofs. They must also extend at least 18 inches beyond the exterior wall of the buildings. The IBC provides specific exceptions for both the vertical and horizontal continuity of the wall. **[Ref. 706.2 and 706.5]**

Fire walls must be fire-resistance rated in accordance with Table 8-6. The rating is based on the occupancy classifications in the building. Fire walls must be constructed of noncombustible materials unless the building is of Type V construction. The openings in a fire wall are limited to 156 square feet in buildings without a fire sprinkler system, and they are limited to 25 percent of the length of the fire wall. The openings must be protected with a fire-rated door or shutter assembly, as shown in Table 8-6. Doors must close by themselves or be automatic closing. They must also latch when closed. Ducts and air transfer openings must be protected with a fire damper. Fire walls that are located on lot lines are called *party walls* and are not permitted to have any openings.

TABLE 8-6 Fire wall ratings and opening ratings (Based on IBC Table 706.4 and 715.4)

Group	Fire Resistance Rating (hours)	Opening Rating
A, B, E, H-4, I, R-1, R-2, U	3[a]	3
F-1, H-3[b], H-5, M, S-1	3	3[c]
H-1, H-2	4[b]	3
F-2, S-2, R-3, R-4	2	1½

a. In Type II and V construction, walls shall be permitted to have a 2-hour fire-resistance rating with 1-½-hour openings.

b. For Group H-1, H-2, or H-3 buildings, also see Sections 415.4 and 415.5.

c. Two doors, each with a fire protection rating of 1½ hours, installed on opposite sides of the same opening in a fire wall, shall be deemed equivalent in fire protection rating to one 3-hour fire door.

TABLE 8-7 Fire door and shutter fire protection of fire barriers (Based on IBC Table 715.4)

Fire Barrier Rating	Fire Door/Shutter Rating
4	3
3	3
2	1½
1	¾[a]

a. Openings in shafts, exit enclosures, and exit passageway walls are required to have a 1-hour fire rating.

Fire barriers

Fire barriers are used to separate different hazards in a building. They can be used to separate different occupancies, incidental uses, and control areas. Fire barriers are also used to protect occupants who are trying to exit the building in an emergency. Unlike fire walls, a fire barrier only needs to extend from the top of the floor/ceiling assembly to the underside of the floor or roof sheathing above. That means that the wall assembly must continue through any concealed spaces, such as spaces above a suspended ceiling. Because it is important for a fire barrier to remain intact and in place during a fire, the construction that supports the wall must have the same fire-resistance rating. **[Ref. 707]**

The rating of a fire barrier is based on the separation that is needed for the particular application. We discussed separated occupancies in Chapter 6. A fire barrier providing a separation between occupancies must have a fire-resistance rating based on IBC Table 508.4. If the fire barrier is separating an incidental use, the rating is based on IBC Table 509. Other applications of a fire barrier require different fire ratings. Therefore, it is important that the user of the IBC pay attention to the specific ratings. Openings in a fire barrier are based on the fire-resistance rating of the wall (see Table 8-7). Doors must close by themselves or be automatic closing. They must also latch when closed. Ducts and air transfer openings in a fire barrier must be protected with a fire damper.

Fire partitions

Fire partitions are most commonly used to separate dwelling units in an apartment or condominium, as well as guest rooms in a hotel or motel (Figure 8-9). The separation between dwelling units or guest rooms provides protection to people sleeping in the adjacent unit from fire that they may not be aware of or have control over. Fire partitions are also used to separate a corridor from adjacent areas of buildings in some occupancies. These corridors provide a smoke-free environment for people to use in getting out of a building during a fire.

FIGURE 8-9 Fire-resistance-rated corridor constructed using fire partitions

Fire partitions are required to be a minimum of 1-hour fire-resistance-rated construction. Door openings located in a separation between dwelling units or guest rooms must have a 1-hour fire-resistance-rating and are not required to be self-closing. Doors in fire-resistance-rated corridors must have a minimum rating of twenty minutes, be self- or automatic closing, and latch. Fire partitions must extend from the floor assembly to either the floor or roof sheathing above, or a fire-resistance-rated floor/ceiling or roof/ceiling assembly. Ducts and air transfer openings must be protected with a fire damper and, in the case of corridors, a smoke damper may be required to be installed. **[Ref. 708]**

Horizontal assemblies

Floor/ceiling and roof/ceiling assemblies are called *horizontal assemblies* in the IBC. A horizontal assembly can serve many purposes. It can serve as an occupancy separation and have a fire-resistance-rating, as required in IBC Table 508.4. It may be required to be rated due to the type of construction as required in IBC Table 601. When a floor/ceiling assembly provides a separation between dwelling units or guest rooms, it must have a 1-hour fire-resistance rating just like a fire partition. Any openings in a horizontal assembly must be protected with a shaft enclosure, as noted earlier. The structure that is supporting the horizontal assembly must be provided with the same fire-resistance-rated protection as the floor or roof that it is supporting. **[Ref. 711]**

Vertical openings

When an opening is placed in a floor assembly, fire and smoke migration from floor to floor can occur. Therefore, the IBC requires that the opening in the floor be protected to limit the spread of fire. Several different are options available in the IBC to provide this protection. The most common option is the construction of a shaft enclosure. Shaft enclosures are discussed later in this chapter.

One of the options where a vertical opening is permitted between two floors is outlined in IBC Section 712.1.8. This option is commonly used for the construction of a two-story lobby in a building. The lobby can be open between the first and second story as long as it meets seven conditions:

1. The opening does not connect more than two stories.
2. The opening does not contain a required stairway or ramp.
3. The opening does not penetrate a horizontal assembly that separates fire areas or smoke barriers that separate smoke compartments.
4. The opening is not concealed within the construction of a wall or a floor-ceiling assembly
5. The opening does not open into a corridor of a Group I and R occupancy.
6. The opening does not open into a corridor on a floor without fire sprinklers.

7. The opening is separated from other floor openings and air transfer openings that serve other floors by construction having the same fire-resistance rating as required for shaft enclosures.

Atriums are also another option that is available to the designer to deal with floor openings. Atriums are typically open to multiple stories in a building. Specific requirements must be met to construct an atrium. They are typically provided with a smoke control system and other fire protection requirements. Atriums are discussed in more detail in Chapter 6. **[Ref. 712.1.6]**

Other common openings between floors are stairways. Vertical floor openings created by unenclosed stairs or ramps need to comply with either Section 1009.2 or 1009.3. These sections address exit access stairways and what conditions need to be met to construct an open stairway between floors.

Shafts

Shaft enclosures are the most common way to protect a vertical opening through a floor-ceiling assembly. These enclosures are constructed as fire barriers. The fire-resistance rating of the fire barriers enclosing a shaft is required to be 2 hours if the shaft connects four or more floors and 1 hour if it only connects two or three floors.

Shafts generally extend from the bottom of the building through the roof. However, in some cases they may not. Therefore, the IBC requires that if the shaft does not go all the way to the bottom of the building, it must comply with one of three requirements:

- The shaft must be enclosed at the bottom with construction having the same fire-resistance rating as the shaft.
- The shaft can terminate in a room having the same fire-resistance rating as the shaft.
- Approved fire dampers can be placed at the floor level where the shaft terminates.

The top of the shaft must either extend to the underside of the roof sheathing or be enclosed at the top with construction having the same fire-resistance rating as the shaft.

Shafts are used for many different purposes in a building. The most common use of a shaft is to enclose mechanical ductwork extending vertically through a building (Figure 8-10). Whenever the duct penetrates the wall of the shaft, the opening must be protected with a combination fire/smoke damper. These dampers close off the opening when fire and/or smoke is present in the duct. Elevator hoistways are also commonly enclosed in a shaft. The elevator doors must be a protected opening to prevent fire from spreading floor to floor as well. Shafts are also used for refuse and laundry chutes. **[Ref. 713]**

FIGURE 8-10 Mechanical ducts penetrating a floor may need to be enclosed in a shaft

FIGURE 8-11 Piping with through-penetration fire stopping

Penetrations

This chapter has discussed openings such as doors being installed in a fire assembly and what type of rating they are required to have. When pipes, tubing, conduit, and cables pass through fire assemblies, they are called *penetrations* and can adversely affect the overall fire-resistance rating. There are two types of penetrations. The first is a through penetration, in which the material passes through the entire assembly. The second is a membrane penetration, in which the material passes through only a single membrane of the fire assembly. Membrane penetrations must comply with the same requirements as through penetrations. Whenever these penetrations occur in fire walls, fire barriers, fire partitions, smoke barriers, and/or horizontal assemblies, they must be protected by a through-penetration fire-stop system (see Figure 8-11).

Firestop material is typically an intumescent type of material. When it is heated, it expands many times its original size. The material is placed around the pipe or cable that penetrates the wall. During a fire, if the pipe or cable burn away, the firestop material fills the hole and prevents the fire from spreading into or through the fire-resistant-rated assembly.

Through-penetration fire-stop systems are required to be tested by an approved testing agency and receive a fire-resistance rating. The systems are given a rating depending on how well they limit the spread through the penetration. The systems are tested for a prescribed time period, which is indicated as an *F rating*. In some applications, such as penetrations in a horizontal assembly, the fire-stop system must have a specific T rating. A *T rating* is the time period that a fire-stop system and the penetrating item limit the rise in temperature on the nonfire side of a fire assembly. The F and T ratings are required to have the same rating as the fire assembly in which the penetration is located. It is important that these assemblies are installed in the same manner as prescribed in the manufacturer's installation instructions. The instructions are based on the specific system that was tested by the testing agency. **[Ref. 714]**

Opening protection

Doors and windows installed in fire-resistance-rated assemblies are required to have a fire protection rating. Fire door assemblies include the door, frame, and all associated hardware. They are tested to a specific standard to determine their fire-resistance rating. Fire door assemblies located in fire-resistance-rated corridors are also required to be designed and tested to provide smoke and draft control. These door assemblies are required to latch so the pressure of a fire does not push open the door. They are also typically required to be installed so they close automatically or on their own so they are not left open under normal conditions.

The rating of doors is based on the assembly in which they are located, as noted in the specific preceding assemblies. Fire window assemblies are required to be tested to a different standard. Fire windows are not permitted to be installed in fire walls or in any fire barriers having a required fire-resistance rating greater than 1 hour. All openings required to be protected must bear a label showing the rating of the assembly. The label must be permanently affixed to the door, window, and associated frame and must be applied by the manufacturer at the factory (see Figure 8-12). **[Ref. 716]**

FIGURE 8-12 Fire-resistance-rated door assembly with label on door and frame

INTERIOR FINISHES

One of the biggest reasons for deaths in buildings is the rapid spread of fire and the toxic chemicals created when finishes on walls and ceilings ignite. Some interior finishes burn faster than others. Therefore, the IBC limits the "flame spread" of materials in buildings. The *flame spread* is the propagation of flame over a surface. Interior finishes are tested by an approved testing agency by placing a sample of the material in a tunnel-like chamber (see Figure 8-13). The agency then ignites the material and measures how fast the flame spreads across the surface. The measurement is a dimensionless number that is compared to a standard material. The test also determines the amount of smoke developed by the burning. The material is then given a "flame-spread index" and "smoke-developed index." The flame-spread indexes of some common materials are shown in Table 8-8. **[Ref. 803]**

The IBC restricts the type of materials to be installed on walls and ceilings based on the flame-spread classification; the occupancy

> **Code Basics**
>
> **Interior Wall and Ceiling Finish Requirements Examples (from IBC Table 803.9)**
>
> Group A-2: Restaurant without Automatic Sprinklers
> - Rooms and enclosed spaces: Class B
> - Corridors: Class A (lobby may be Class B)
> - Exit enclosures: Class A
>
> Group A-2: Restaurant, Sprinklered
> - Rooms and enclosed spaces: Class C
> - Corridors: Class B
> - Exit enclosures: Class B

FIGURE 8-13 Steiner Tunnel for testing flame spread of interior finishes

TABLE 8-8 Flame spread of common materials

Material	Flame Spread
Maple	104
Red or White Oak	100
Douglas-fir	90
Western Red Cedar	69
Plywood paneling (untreated)	75 to 275
Plywood paneling (treated)	100
Gypsum board	10 to 25
Concrete or brick	0
Carpeting	10 to 600

classification; the location where the materials will be installed (rooms and enclosed spaces, corridors, and exit enclosures and exit passageways); and whether or not the building is provided with a fire sprinkler system. There are three levels of flame-spread classification:

Class A: Flame-spread index 0–25; smoke-developed index 0–450
Class B: Flame-spread index 26–75; smoke-developed index 0–450
Class C: Flame-spread index 76–200; smoke-developed index 0–450

The flame-spread classification requirements are outlined in IBC Table 803.9.

> **Code Basics**
>
> **Fireblocking Materials**
>
> 1. 2-inch nominal lumber
> 2. Two thicknesses of 1-inch nominal lumber with broken lap joints
> 3. One thickness of 23/32-inch wood structural panels with joints backed by 23/32-inch wood structural panels
> 4. One thickness of ¾ inch particleboard with joints backed by ¾-inch particleboard
> 5. ½-inch gypsum board
> 6. ¼-inch cement-based millboard
> 7. Batts or blankets of mineral wool, mineral fiber, or other approved materials installed in such a manner as to be securely retained in place ●

STOPPING HEATED AIR AND FIRE MOVEMENT

When a fire occurs in a building, hot gases, smoke, and flames spread through paths of least resistance. There are spaces in a building constructed of combustible materials where fire can travel undetected. These areas are referred to as *concealed spaces*. If the fire gets into these areas, it is very difficult for firefighters to see where the fire is and extinguish the flames. In many cases, fire can spread through these concealed spaces and pose serious danger to the firefighters due to collapse of the structure. The IBC controls the spread of fire in concealed spaces in two ways, fireblocking, and draftstopping.

Fireblocking

Fireblocking is used to cut off concealed draft openings within a wall and between a wall and floor or attic spaces (Figure 8-14). It also provides a barrier between floors and between the top floor and the attic. The IBC specifies the type of materials permitted to be used as fireblocking. Some of the materials are noncombustible and others are combustible. It is important to remember that the intent is to stop the draft within a concealed space. If the movement of heated air is stopped within these spaces, the spread of fire will be severely restricted. **[Ref. 718.2]**

Sections 718.2.2 through 718.1.7 of the IBC outline where fireblocking is required. Locations where fireblocking is required include:

- Concealed wall spaces at the ceiling and floor levels as well as horizontally at intervals of no more than 10 feet

- Interconnections between concealed wall spaces and concealed horizontal spaces in a floor or roof assembly (Examples of these spaces include soffits, drop ceilings, and cove ceilings.)
- Concealed spaces between stair stringers at the top and bottom of the run of stairs
- Open spaces around vents, pipes, ducts, chimneys, and fireplaces at ceiling and floor levels with specific tested material
- Concealed spaces of exterior wall finish and other combustible exterior architectural elements at intervals of 20 feet, so that no open space exceeds 100 square feet.
- Concealed spaces formed by wood floor sleepers on masonry or concrete fire-resistance-rated floors

FIGURE 8-14 Plywood fireblocking to stop draft from going from wall cavity to soffit and then into the floor space above

Draftstopping

Draftstopping is used in large attics and concealed floor spaces in combustible construction. In very large buildings, these spaces can be rather large. If a fire gets into these areas, it can result in a very large fire loss and potentially loss of life, including firefighters (Figure 8-15). The materials used for draftstopping include ½-inch gypsum board, ⅜-inch wood structural panel, ⅜-inch particle board, 1-inch nominal lumber, cement fiberboard, and batts or blankets of mineral wool or glass fiber.

FIGURE 8-15 Draftstopping in floor/ceiling assembly

In Group R occupancies, draftstopping is required to be installed within a floor assembly above and in line with the dwelling unit or guest room separations. In other occupancies, it is required in the floor assembly so that the area of the concealed space does not exceed 1,000 square feet. If fire sprinklers are installed within the floor cavity, the draftstopping is not necessary. If fire gets into the concealed space, the fire sprinklers will activate and stop the fire from spreading through the space.

Attics, mansards, overhangs, or other concealed roof spaces in Group R-1 and Group R-2 occupancies must also be provided with draftstopping above and in line with the dwelling unit or guest room separations. Other occupancies must have draftstopping in attics and concealed roof spaces so that the horizontal area of the concealed space does not exceed 3,000 square feet. When fire sprinklers are installed in these spaces, the draftstopping requirements are eliminated. **[Ref. 718.3 and 718.4]**

CHAPTER

9 Active Fire Protection

© International Code Council

Passive fire protection was discussed in Chapter 8. It incorporates the fire protection that is built into the building. Active fire protection, on the other hand, is a system that is used to notify the occupants of a fire or to control the flames and/or smoke. The most common types of active protection are automatic fire sprinklers and fire alarm systems. This chapter discusses fire sprinklers and other types of systems that are considered active fire protection systems.

AUTOMATIC SPRINKLER SYSTEMS

Automatic sprinklers work by spraying water on the fire and either extinguishing or suppressing it until the fire department can respond and extinguish the fire. Hollywood movies incorrectly portray fire sprinklers' operation. Filmmakers seem to want you to think that fire sprinklers go off in an entire building at the same time. This cannot be any further from the truth. When a fire occurs in a sprinklered building, only the sprinklers activated by the fire disperse water on the fire. That can be anywhere from one to four sprinklers, depending on the design of the system.

A fire sprinkler is a device with an opening that is closed with a plug held in place by a fusible link or bulb filled with liquid. A fusible link is two pieces of metal soldered together. When the temperature around the sprinkler head reaches a specific temperature, the link or bulb breaks and the plug pops out of the opening. The water flows through the opening and hits a deflector below that sprays the water outward toward the fire. In many cases, the fire can be contained with only one sprinkler head. If the fire overruns a single head, additional heads are then activated (see Figure 9-1, *a–d*).

Automatic sprinkler systems are typically connected to the public water supply. If there is no public water system available, a storage tank and pump system are provided. The system is designed so the pressure of the water system is maintained to ensure that the minimum number of heads can be used to control the fire. When the fire department arrives at the fire scene, firefighters connect their pumper to a fire hydrant and then to a fire department connection on the outside of the building (see

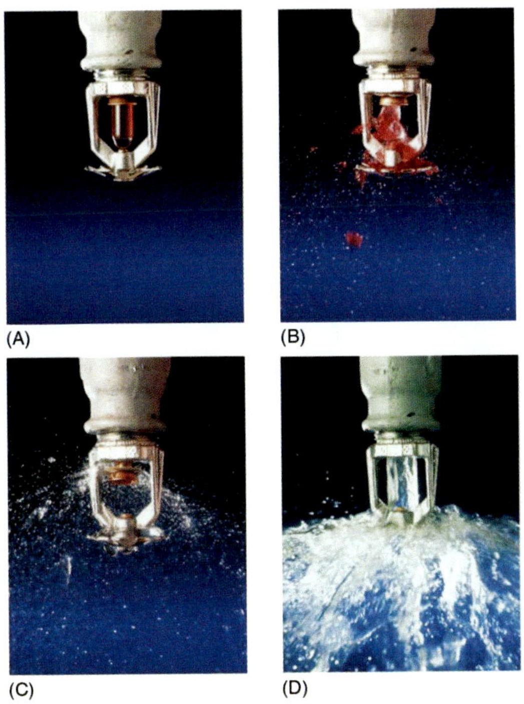

FIGURE 9-1 Activation of sprinkler head

FIGURE 9-2 Exterior sprinkler supply. Fire department connection located on right side halfway up

Figure 9-2). The fire department connection is tied to the sprinkler system piping. Firefighters can then increase the pressure on the system to assist in controlling the fire.

There are two primary reasons that automatic sprinklers are installed in buildings. The first is when the IBC requires a system to be installed based on specific conditions. The second is to take advantage of credits given by the IBC for the installation of an automatic sprinkler system. In some cases, sprinkler systems are installed for both reasons. In most cases, the credits can be taken even when the systems are required to be installed in a building.

Where are automatic sprinklers required?

The IBC requires fire sprinklers in buildings based on the *fire area.* The fire area is a compartment in a building that is surrounded by exterior walls, fire walls, fire barriers, or horizontal assemblies (see Figure 9-3). If the building has an outdoor area covered by a roof or floor above, the covered area is considered to be a part of the fire area as well. A building by itself can be considered a single fire area. When a building is separated by a fire wall to create two separate buildings, each of the buildings is considered a separate fire area as well. The fire-resistance rating of the fire walls creating separate fire areas must comply with Table 706.4, just like any other fire wall. If the building is divided by fire barriers and horizontal assemblies, it could have multiple fire areas. The fire-resistance ratings of fire barriers and horizontal assemblies creating separate fire areas are shown in Table 9-1. **[Ref. 707.3.10]**

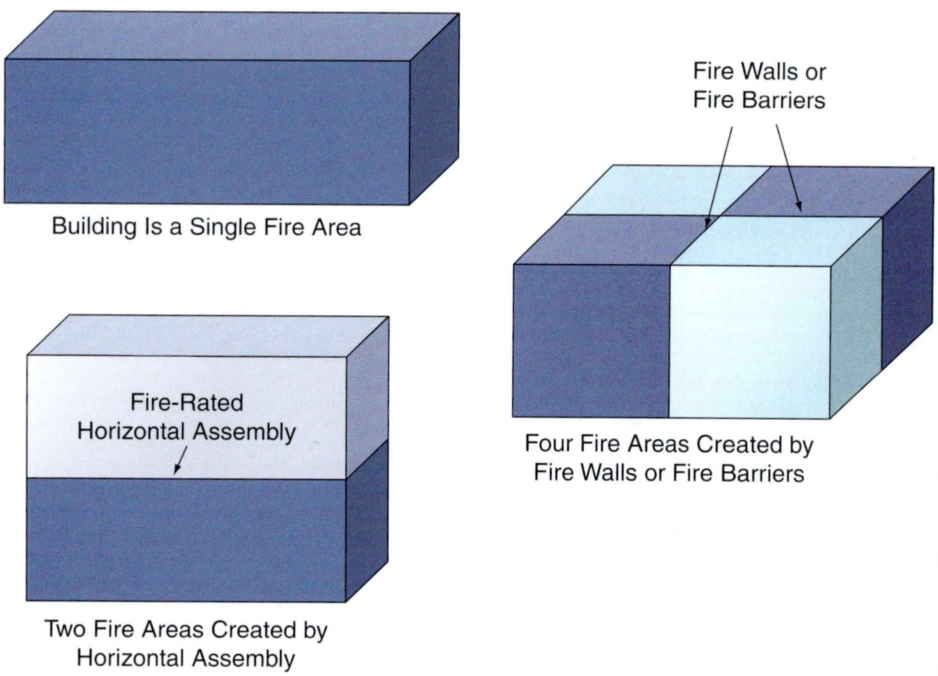

FIGURE 9-3 Fire area examples

Once the fire area of a building is determined, the IBC specifies which occupancies require fire sprinklers. Section 903 lists the

TABLE 9-1 Fire ratings for fire barriers and horizontal assemblies between fire areas (Based on IBC Table 707.3.9)

Occupancy Group	Fire Resistance Rating (hours)
H-1, H-2	4
F-1, H-3, S-1	3
A, B, E, F-2, H-4, H-5, I, M, R, S-2	2
U	1

TABLE 9-2 Examples of occupancy groups that require fire sprinklers (Based on a portion of IBC section 903)

Occupancy Group	Fire Sprinklers Required Where One of the Following Conditions Exist
Group A-2 Restaurant/Night Club	1. The fire area exceeds 5,000 square feet;
	2. The fire area has an occupant load of 100 or more; or
	3. The fire area is located on a floor other than a level of exit discharge serving such occupancies.
Group M Mercantile	1. Where a Group M *fire area* exceeds 12,000 square feet;
	2. Where a Group M *fire area* is located more than three stories above *grade plane;*
	3. Where the combined area of all Group M *fire areas* on all floors, including any mezzanines, exceeds 24,000 square feet; or
	4. Where a Group M occupancy is used for the display and sale of upholstered furniture that exceeds 5,000 square feet.
Group R Residential	Sprinklers required throughout all buildings containing a Group R fire area.

occupancies that are required to be provided with fire sprinklers. The requirements are typically based on the area of the fire area, the location of the fire area in relation to the exterior grade around a building, the occupant load of the building, and sometimes specific uses within an occupancy group. Table 9-2 provides a few examples of the sprinkler requirements in more common occupancies. **[Ref. 903.2]**

In some cases, buildings are required to be provided with fire sprinklers regardless of their occupancy classification. Buildings that are not provided with openings in the exterior wall so that firefighters can access the building to extinguish the fire must be provided with fire sprinklers. Any building that is over 1,500 square feet in area and does not have a minimum amount of openings falls into this category. The minimum openings located below ground must be accessed by an exterior stairway or ramp. The openings must be located in each 50 linear feet of the exterior wall on at least one side. Openings above ground must be a minimum of 20 square feet located each 50 linear feet

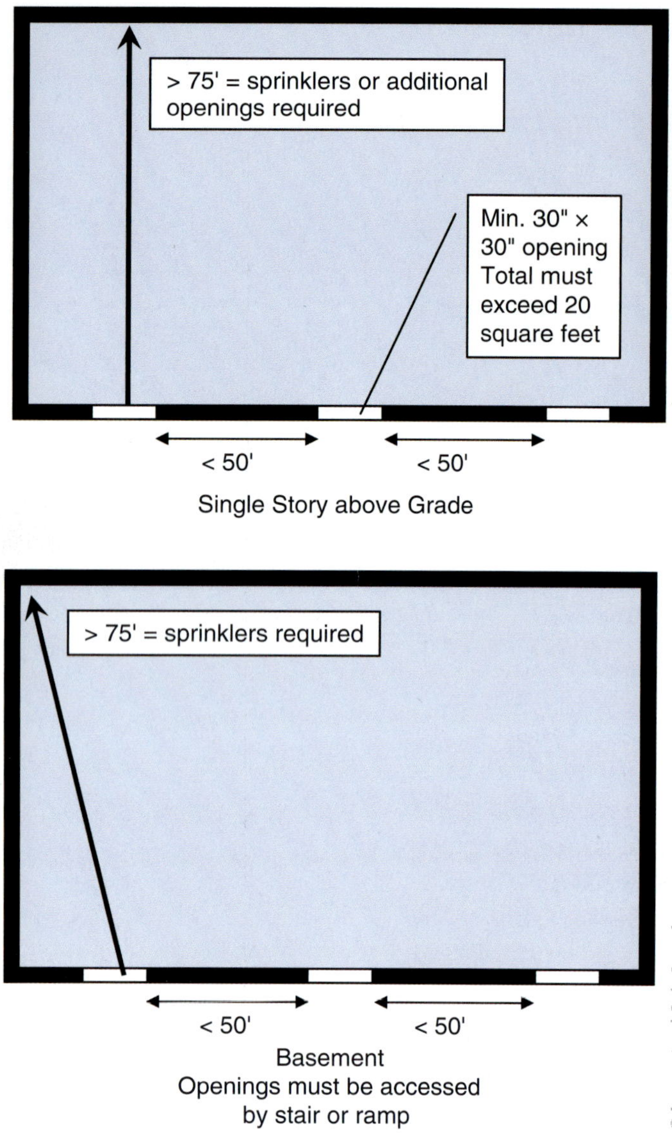

FIGURE 9-4 Automatic sprinkler requirements based on exterior openings

along the exterior wall on at least one side of the building. The openings must be at least 30 inches in dimension to allow firefighter access into the building. In addition, openings are not permitted to be obstructed in any way that may restrict firefighting or rescues.

When the minimum openings are located on only one side of a building, firefighter access is restricted to only one side of the building. When the firefighters access a building with a fire hose, they can only go as far as the length of the hose will permit them. Therefore, the IBC states that when the only openings in a story are located on one side of the building, and the opposite wall of a story is more than 75 feet from such openings, either the building must be provided with fire sprinklers or additional openings as we have described must be provided on at least one other side of the building. For basements, if any portion of the basement is farther than 75 feet from an opening, or when walls, partitions, or other obstructions are installed that restrict the application of water from hose streams, fire sprinklers must be provided in the basement (see Figure 9-4). **[Ref. 903.2.11.1]**

Because it is difficult to fight fires in tall buildings from the outside, fire sprinklers are required in buildings having a floor level with more than thirty occupants and located 55 feet or more above the lowest level of fire department vehicle access. This applies to all occupancy groups with the exception of open parking structures, Group F-2 low-hazard factories, and airport control towers. Fire sprinklers are also required in rubbish and linen chutes and in their terminal rooms. When a chute extends through three or more floors, sprinklers must be installed within the chutes at alternate floors. **[Ref. 903.2.11]**

Sprinkler credits

When fire sprinklers are installed in a building, the IBC gives credit for the added protection the system provides. These credits or modifications are generally permitted even when the system is required to be installed throughout by Section 903 of the code. The installation of fire sprinklers can increase the permitted area and height of the building as discussed earlier in this book. We also discussed the increase in the maximum allowable quantities of hazardous materials when fire sprinklers are installed throughout a building. Fire sprinklers can be used to modify many other requirements of the code in several other different ways.

The modifications permitted are sometimes limited to the type of sprinkler system to be installed. Three different standards are used in the design of automatic sprinkler systems. The National Fire Protection Association (NFPA) develops these standards. NFPA Standard 13 is used to design and install automatic sprinkler systems in most buildings. This type of system is primarily designed to protect the building and occupants from fire. The standard requires that all portions of the building be protected by sprinklers. This includes concealed combustible spaces like open-frame floor systems and attics.

FIGURE 9-5 **NFPA 13R Sprinkler Systems are limited to residential occupancies with a maximum of four stories**

An automatic sprinkler system installed in accordance with NFPA Standard 13R is designed for residential occupancies up to and including four stories in height. This system is designed to detect and control fires in residential occupancies. Its primary purpose is to protect the occupants of the building. NFPA 13R systems are not as restrictive as NFPA 13 systems. The design requirements do not require concealed combustible spaces to be provided with sprinklers. It also allows bathrooms less than 50-square feet and small closets to be constructed without having to be provided with sprinklers. Both NFPA 13 and NFPA 13R systems are supplied with a separate water supply from the building domestic water system (see Figure 9-5).

NFPA Standard 13D is used for sprinkler systems installed in one- and two-family dwellings and manufactured homes. This system is less restrictive than the NFPA 13R systems. It is designed to protect the occupants within a dwelling unit by suppressing the fire long enough for them to get out. In many cases, the systems extinguish the fire before the fire department arrives. Sprinkler heads are not required in concealed combustible spaces, small bathrooms, and small closets. An NFPA 13 D system is permitted to be connected to the domestic water supply.

The fire sprinkler modifications in the IBC require a system installed according to either NFPA 13 or NFPA 13R, or both. It is important that the user of the code pay close attention to the exceptions and requirements permitting an automatic sprinkler system to be substituted for an IBC requirement. For example, the allowable area of the building can be increased when an NFPA 13 system is installed but not when an NFPA 13R system is installed. However, the height of a residential building can be increased one story and 20 feet if either type of system is installed throughout the building when either a NFPA 13 or 13R system is installed throughout the building. NFPA 13D systems are not permitted to be used for any modifications or substitutions. **[Ref. 903.3.1]**

FIGURE 9-6 Standpipe located in parking garage

FIRE STANDPIPES

Fire standpipes are an extension of the fire pumper into a building. When a building reaches a certain size or height, a firefighter cannot carry enough fire hose to access the fire. Therefore, the IBC requires that fire standpipes be installed. Standpipes are a series of pipes that run from the fire department connection on the outside of the building to fire department hose connections inside the building. When the fire department responds to a fire in a building, firefighters connect the fire pumper between a fire hydrant and the fire department connection. They fill the standpipe system with water so they can connect fire hoses to the hose connections and fight the fire from within the building. Standpipe systems must be designed and installed in accordance with NFPA Standard 14 (see Figure 9-6). **[Ref. 905.2]**

The two major types of standpipe systems are wet and dry. Wet standpipe systems are generally connected to the same water supply as the automatic sprinkler system. This means that there is water in the standpipe at all times. The fire department can still connect to the fire department connection to provide additional amounts of water and pressure. Wet standpipe systems are then divided into automatic wet and manual wet systems. Automatic wet standpipes are typically connected to a fire pump that can provide the required demand of water and pressure without the fire department connecting to it. A manual wet standpipe system cannot meet the water and pressure demands and must be pressurized by the fire pumper.

Dry standpipes are not filled with water. The most common type of dry system is a manual dry standpipe. These systems require water from the fire department to work. The fire department must connect to the fire department connection and deliver water to the system. An automatic dry standpipe is normally filled with pressurized air. When a firefighter connects a hose to the fire hose connection and opens the valve, an automatic valve admits water into the system. Dry standpipes are only permitted to be installed when they are subject to freezing, such as in a parking garage in a cold climate.

Standpipe systems are also given one of three class designations: Class I, Class II, or Class III. A Class I system has 2½ inch hose connections and is designed to be used by firefighters trained to handle heavy streams of water. A Class II system is provided with 1½ inch hose connections. It is designed to be used by the building occupants or firefighters during the initial response. Many fire departments prefer that occupants not use these systems unless they have been trained in the proper use of the hoses. They would rather have people evacuate the building and let the firefighters extinguish the flames. A Class III standpipe system is provided with both a 2½ inch and

1½ inch hose connection (see Figure 9-7). This is usually done with a reducing fitting. These systems are designed to be used by both the building occupants and the fire department. [Ref. 902]

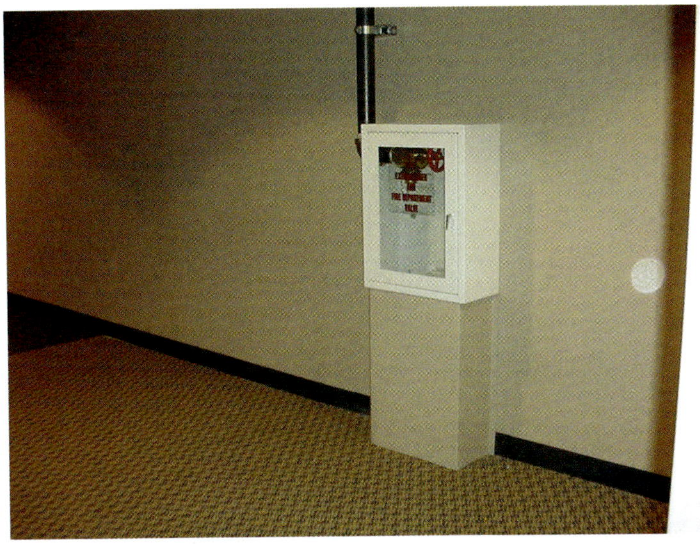

FIGURE 9-7 Hose cabinet with Class III standpipe connection

Where are standpipes required?

As already mentioned, standpipes are needed in buildings that are too large to use a fire hose from the outside to fight a fire. Therefore, Section 905 of the IBC designates when buildings must be provided with standpipes and the type of standpipe required. Because standpipes and automatic sprinkler systems are used for the same purpose, they are permitted to be combined.

The maximum height that a firefighter can typically pull a fire hose into and up the stairs of a building is 30 feet. Therefore, the IBC requires that standpipes be installed in buildings in which the floor level of the highest story is more than 30 feet above or below the lowest level of fire department vehicle access (see Figure 9-8). Class III standpipe systems must be installed unless the building is provided with an automatic sprinkler system where a Class I system can be installed. Open parking garages are also permitted to install Class I systems under specific conditions. [Ref. 905.3.1]

FIGURE 9-8 Buildings with an occupied floor more than 50 feet above fire department vehicle access must be provided with fire standpipes

Standpipes are also required to be installed in very large buildings. Covered mall buildings must be provided with standpipes if they have a floor that is more than 30 feet above the lowest level of fire department vehicle access. If a floor does not meet this criterion, Class I hose connections must be provided at specific locations in the covered mall building.

Class I hose connections must be located in areas that firefighters can access and make a safe connection. The most common location for these connections is within an enclosed stairway. They must be located at an intermediate floor level landing to provide access to two floors and provide some separation between the fire and the connection. Connections are also required on each side of a horizontal exit, inside exit passageways, and adjacent to entrances to a covered mall building. Hose connections are also required at the roof if the slope of the roof is less than four units vertical in twelve units horizontal. [Ref. 905.4]

Class II hose connections must be located in such a manner that every portion of a building is within 30 feet of a nozzle attached to 100 feet of hose. The distance is measured as the hose would be pulled through the building. So, it needs to extend around walls and into every room. Class III hose connections must be located as required for both Class I and Class II systems. [Ref. 905.5]

FIRE ALARMS

A fire alarm system provides early detection of a fire and warns the occupants of a building of the danger. Early detection of fire provides occupants enough warning to get out of a burning building before the situation gets to a point at which they could become injured or killed. Two types of fire alarm systems are required by Section 907.2 of the IBC. A manual fire alarm system has manual fire alarm boxes that occupants must activate to initiate an alarm signal. These boxes are sometimes called *alarm pull stations* (see Figure 9-9).

FIGURE 9-9 Fire alarm pull station next to stair tactile sign

The second type of fire alarm system is an automatic smoke-detection system. These systems use smoke detectors for the protection of an area. When a smoke detector senses smoke, it initiates the alarm automatically and provides early warning of fire. In areas in which the ambient temperatures prevent the use of smoke detectors, approved automatic heat detectors can be installed instead.

Where are fire alarms required?

Fire alarm systems are typically required based on the occupancy classification. In some occupancy groups, only a manual fire alarm system is required. In others, an automatic smoke-detection system is required. It is important that the user of the IBC pay attention to the specific requirements for the occupancy under consideration. Regardless of the type of system installed, at least one manual fire alarm box must be provided in a location approved by the local authority having jurisdiction. The exception to this rule is when a fire alarm system is dedicated to elevator recall control, or in Group R-2 occupancies. [Ref. 907.2]

Manual fire alarm systems are required in the following locations:

- Group A occupancies having an occupant load of three hundred or more
- Group B occupancies having more than one hundred persons above the level of exit discharge; when the combined occupant load of all floors is five hundred or more; and when a Group B ambulatory health-care facility is located in the building
- Group E occupancies with an occupant load of 50 or more
- Group F occupancies that are two or more stories in height and the combined occupant load above or below the lowest level of exit discharge is five hundred or more
- Group H-5 occupancies and in occupancies used for the manufacture of organic coatings

- Group I occupancies
- Group M occupancies when the combined occupant load on all floors is five hundred or more or when the occupant load is more than one hundred above or below the lowest level of exit discharge
- Aerosol storage areas
- Lumber, wood structural panel, and veneer mills
- Deep underground buildings

In many cases, the IBC permits the elimination of the manual fire alarm pull stations when the building is provided with an automatic sprinkler system. However, the overall fire alarm system with the proper notification devices must be provided. Most of these exceptions require that the activation of a sprinkler head trigger a flow switch that is connected to occupant notification devices. These devices provide the early warning of a fire in the building for the people.

Automatic smoke-detection systems are required in the following locations:

- Group B ambulatory health-care facilities
- Group H occupancies with highly toxic gases, organic peroxides, and oxidizers
- Group I-1 occupancies in corridors, waiting areas open to corridors, and habitable spaces other than sleeping units and kitchens
- Group I-2 occupancies in corridors in nursing homes, detoxification facilities, and spaces permitted to be open to corridors
- Group I-3 occupancies throughout resident housing areas
- Group R-1 occupancies throughout all interior corridors serving sleeping units
- Special amusement buildings
- High-rise buildings
- Atriums that connect more than two stories
- High-piled combustible storage areas
- Underground buildings with smoke control
- Airport traffic control towers
- Battery rooms

When a manual fire alarm pull station is activated, or an automatic smoke detector or automatic sprinkler system activates, the occupant notification portion of the fire alarm system must be initiated. Three different notification systems are used to notify the building occupants of a fire. Audible notification appliances emit a distinctive sound that cannot be used for any purpose other than that of a fire alarm. The appliance must emit an alarm at a sound level of fifteen decibels (dBA) above the average ambient sound level or five dBA above the maximum sound level having a duration of at least sixty seconds, whichever is greater, in every occupiable space of a building. **[Ref. 907.5.2.1]**

Emergency voice/alarm communication systems automatically sound an alert tone followed by recorded voice instructions giving information and directions for general or staged evacuation. The fire department has the ability to override the system and provide more specific directions

FIGURE 9-10 Combination audible/visual fire alarm notification device

if needed in a fire event. Emergency voice/alarm systems must be provided with an emergency power source to ensure the system works with power being disconnected to the building. **[Ref. 907.5.2.2]**

Visible alarms are typically strobe lights attached to the audible alarm device. They are provided in a building for occupants that may be hearing impaired. They must be located in public areas and common areas. They must also be provided in Group I-1 and Group R-1 dwelling units or sleeping units that may be occupied by disabled persons. Both the building fire alarm system and the smoke alarm system in the units must activate the visible alarm notification appliance (see Figure 9-10). **[Ref. 907.5.2.3]**

Fire alarm systems must be installed in accordance with NFPA Standard 72. They must be monitored by an approved supervising station in accordance with this standard. When the system installation is complete, it must be tested to confirm that it will work in the manner that it was designed. **[Ref. 907.6]**

Single-station smoke alarms

A single-station smoke alarm is a device that includes the detector, control equipment, and alarm sounding device in a single unit. The smoke alarms in most homes and apartment dwelling units are considered single-station smoke alarms. When two or more of these devices are combined, the system is classified as a multiple-station smoke alarm. When one device senses smoke, all of the devices on the system sound an alarm. Single- and multiple-station smoke alarms are not connected to the rest of the building fire alarm system. They only notify the occupants within the space that they are located.

FIGURE 9-11 Single-station smoke alarm located in hotel room

Single- or multiple-station smoke alarms are required to be installed within Group R occupancies. In Group R-1 occupancies, they must be installed in the sleeping areas, in every room in the egress path from the sleeping room to the door leading out of the sleeping unit, and in each story of the sleeping unit including basements. In Groups R-2, R-3, R-4, and I-1 occupancies, the smoke alarms must be located in each sleeping room, on the ceiling or wall immediately outside each separate sleeping area, and in each story within a dwelling unit including basements (see Figure 9-11). **[Ref. 907.2.11]**

The smoke alarm system must be provided with primary power from the building wiring system served by a commercial source. The system must also be provided with battery backup to ensure that the alarms will work during a power outage. The alarms must be interconnected so that the activation of one alarm will activate all of the alarms in a sleeping or dwelling unit.

CONTROLLING SMOKE

The migration of smoke throughout a building can have devastating effects on the occupants. The IBC requires control of this smoke migration in specific types of buildings to limit the amount of smoke within the means of egress. The intent of a smoke-control system is to provide a tenable environment for the evacuation or relocation of the people in the building. These systems are not intended for the preservation of contents, restoration of operations, or assistance in fire suppression or overhaul activities. There are three different methods of providing smoke control in a building:

1. The prescriptive provisions of the code regarding smoke barriers, smoke partitions, horizontal smoke barriers, and smoke compartments
2. Smoke-protected assembly seating, which typically is used for exterior assembly seating and large arenas
3. Smoke-control systems as established in IBC Section 909

Smoke-control methods

The design of pressurization, airflow design, and exhaust method smoke-control system as established in IBC Section 909 can be a very complex and difficult system to design, review, and inspect. It is recommended that people having experience in the area of smoke control be involved in the process of developing smoke-control systems. These types of smoke-control systems are typically used in atriums open to three or more stories, malls having atriums, underground buildings, and large arenas.

The smoke barrier construction method is more prescriptive in its design. Smoke barriers are 1-hour fire-resistance-rated walls and/or floors that form a continuous membrane designed and constructed to restrict the movement of smoke. Smoke barriers are commonly used in Group I-2, hospitals, and Group I-3, jails/prisons. The intent of the smoke barriers is to allow patients to be moved from one side of a smoke barrier to another smoke compartment away from the fire. This is called "protecting in place." This gives the patients at least 1 hour of protection and allows the fire department more time to evacuate patients from the building. Smoke barriers must comply with Section 709, as well as Section 909.5, of the IBC when used for smoke control purposes. Section 709 describes the construction of the assembly based on the fire-resistance-rating requirements. Section 909.5 regulates the leakage area of the assemblies to limit the spread of fire (see Figure 9-12).

FIGURE 9-12 Cross-corridor doors located at smoke barrier in hospital

Fire and smoke are a large cause of death and injuries in a building. By providing a level of fire protection from passive and active fire protection systems, we can reduce the number of deaths and injuries. The specific systems must be installed in accordance with the IBC, standards, and/or manufacturer's installation requirements to ensure that they perform as they are designed. The reduction of the number of deaths over time has shown that these requirements are accomplishing their ultimate goal of saving lives. **[Ref. 909]**

CARBON MONOXIDE DETECTORS

The combination of the installation of fuel-burning appliances and improved energy codes has created much tighter buildings. Therefore, the IBC requires that Group I and R occupancies be provided with carbon monoxide alarms when they contain fuel-burning appliances or the building includes an attached garage. The alarms must be listed in accordance with UL 2034 and installed in accordance with NFPA 720.

NFPA 720 requires that the alarms must be installed in the followings locations:

- Outside of each separate dwelling-unit sleeping area in the immediate vicinity of the bedrooms
- On every level of a dwelling unit that can be occupied, including basements but excluding attics and crawl spaces

PART IV

Life Safety

CHAPTER 10
Getting People Out

© International Code Council

Whenever an emergency situation happens in a building, it is important to evacuate people in a safe and efficient manner. The design of a building must enable people to exit the building and reach a safe place based on the hazards that may be present based on the occupancy classification. In institutional occupancies, it may not be possible or desirable to get occupants out of a building. Therefore, the patients or prisoners are moved to other portions of the building that are protected from the hazard. In some cases, the hazards may be the materials in the building; in others, it may be the number of people in a building. This chapter discusses how the *International Building Code* accomplishes the goal of getting people out.

DESIGN OF EXIT SYSTEMS

Designing the exit system in a building is like designing a plumbing system. When designing a plumbing system, you need to know how much stuff is going down the pipe. Once you have this information, you can design the number of pipes, the size of the pipes, and the length of the pipes. Plumbing systems need to provide an efficient path for the materials to move through the piping system. If the system is blocked, the materials can no longer pass through. The exit system is very similar in the design. The first step is to determine how much stuff (people) are in the building. Based on this, you can determine how many exits, the width of the exits, and the maximum distance that a person has to travel to get to an exit. The IBC calls the exit system in a building the means of egress (see Figure 10-1).

FIGURE 10-1 People exiting an upper level and going out the exterior exit door

Number of people

The IBC determines the potential number of people in a building by calculating the design occupant load. The occupant load is a design factor to determine the design of the means of egress. It is not necessarily the actual number of people that may be in a building. When buildings or areas are not provided with fixed seating, the occupant load is based on a calculation using the floor area of a space. In buildings or spaces that have fixed seating, the occupant load is based on the number of seats. When a building has two or more uses, the design of the means of egress applies to each portion of the building based on the use of that space. Outdoor areas must also be provided with a means of egress in accordance with the IBC. **[Ref. 1004]**

Areas without fixed seating

The design occupant load in areas without fixed seating is calculated by taking the floor area and dividing it by the occupant load factors in IBC Table 1004.1.1. A sampling of this table is shown in Table 10-1. The table has two different types of floor area conditions, gross and net. Gross floor area is the entire floor area within the exterior walls, exclusive of vent shafts and courts. This area includes all areas of the space including corridors, stairways, closets, and similar areas. In areas that do not have exterior walls, the usable area under the horizontal projection of the roof or floor above is used to calculate the occupant load. Net floor area is the actual occupied area within the exterior walls. In the case of net floor area, the designer can deduct the unoccupied accessory areas such as corridors, stairways, toilet rooms, mechanical rooms, and closets. Both the gross and net area factors take into account any furniture, fixtures, equipment, or so on that are expected in the space under consideration. **[Ref. 1004.1.1]**

TABLE 10-1 Occupant load design factors (Based on IBC Table 1004.1.1)

Function of Space	Floor Area in sq. ft. per Occupant
Assembly without fixed seats	
Concentrated (chairs only—not fixed)	7 net
Standing space	5 net
Unconcentrated (tables and chairs)	15 net
Business areas	100 gross
Kitchens, commercial	200 gross
Mercantile	
Basements and grade floor areas	30 gross
Areas on other floors	60 gross
Storage, stock, shipping areas	300 gross
Residential	200 gross
Warehouses	500 gross

Areas with fixed seating

The occupant load for areas where the seats are fixed to the floor is based on the number of fixed seats that are installed in a space. In those areas that have both fixed seating and areas without fixed seating, the occupant load is calculated based on the floor areas of each type of seating noted previously. When a space has fixed seating without dividing arms, such as church pews, the number of seats is based on one person for each 18 inches of seating length. Seating booths in restaurants are measured at 24 inches per person. The measurement is taken at the backrest of the seating booth. **[Ref. 1004.4]**

EXAMPLE

Occupant load calculation

Given:

Restaurant as shown
5,000 sq. ft. of area of tables and chairs
8 booths (16 benches) that are 6 ft wide each
800 sq. ft. kitchen

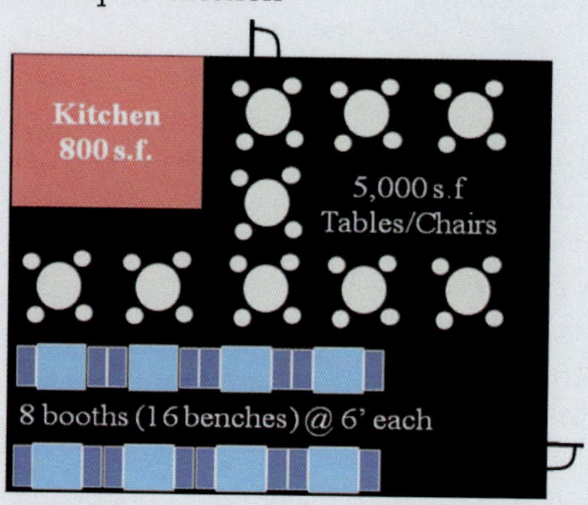

Kitchen	800/200 = 4
Dining Area	5,000/15 = 333
Booths	6'/24" = 3/bench
	3 × 16 = 48
Total	385

Number of ways out

As the occupant load increases, the number of ways out of a building needs to be increased to handle the capacity and to give the people additional options to get out. The IBC outlines the number of exit paths out of a room or space and then the number of exits out of a building, just like the branch lines of a plumbing system. The number of exits is determined by the occupant load in the space or building. Each space is evaluated individually to determine the number of exits needed. To determine the number of exits from a building, all of the occupant loads for each of individual areas are added together to determine the total number of exits required for the building.

Exits from spaces

Rooms and spaces are located in what the IBC calls the *exit access*. The exit access is the first of three parts of the means of egress. The majority

TABLE 10-2 Spaces with one exit or exit access doorway (Based on IBC Table 1015.1)

Occupancy	Maximum occupant load
A, B, E[a], F, M, U	49
H-1, H-2, H-3	3
H-4, H-5, I-1, I-3, I-4, R	10
S	29

a. Day care maximum occupant load is 10.

of the floor area in a building falls within the exit access. It is essentially the unprotected part of the means of egress. When evaluating individual rooms or spaces, the occupant load must be calculated and the number of egress openings determined. In some cases, the occupant load may be relatively low, but the space is of such size that additional egress openings are required. This is discussed in more detail later in this chapter.

Section 1015 of the IBC outlines the requirements for the number of exits or exit access doorways from spaces. The first requirement states that spaces that have an occupant load greater than the values in Table 10-2 must have a minimum of two exits or exit access doorways. For example, a 3,000 square foot office tenant space having an occupant load of 30 (3,000 ÷ 100 = 30) only needs one exit or exit access doorway from the tenant space. However, if the tenant space is 6,000 square feet in area, the occupant load would be 60 (6,000 ÷ 100 = 60), and a minimum of two exit access doorways must be provided.

In very large spaces, additional exits or exit access doorways may be required. When the design occupant load is 501 to 1,000, a minimum of three exits or exit access doorways is required. If the design occupant load exceeds 1,000, at least four exits or exit access doorways will be required. For example, a large ballroom in a hotel might have a floor area of 10,000 square feet. The occupant load factor for the space would be 7 square feet per occupant. The calculated occupant load would be 1,429. This would require a minimum four means of egress from the ballroom. Each means of egress may consist of one or more exits or exit access doorways based on the required egress width as discussed later in this chapter. This is the maximum number exits or exit access doorways required by the IBC. **[Ref. 1015.1.1]**

In some cases, a space may require at least two exits or exit access doors not due to the calculated occupant load but because the space is somewhat large. A large space could create a dangerous dead-end condition in which an occupant may become trapped in the space. The code calls this dead-end condition the "common path of egress travel." The common path of egress travel is the distance that a person has to travel within a space before he or she gets to a point that he or she has a choice of two directions to get to two separate protected exits or the exterior doors of a building. The IBC states that when the length of the common path of egress travel exceeds a specific distance, two exit access doors or exits must be provided from that space. The distance is measured along a natural and unobstructed path of egress travel (see Figure 10-2). **[Ref. 1014.3]**

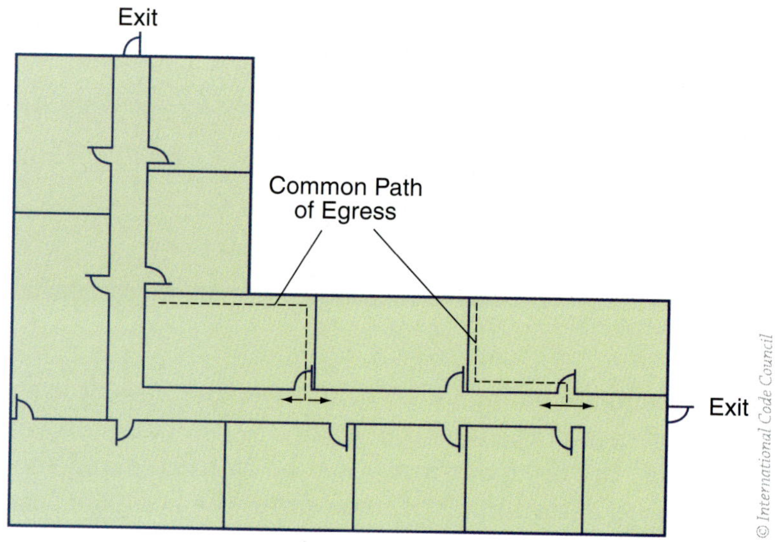

FIGURE 10-2 Common path of egress travel measurement

The maximum common paths of egress travel before two exits or exit access doors are provided are as follows:

- 75 feet in occupancies other than Group H-1, Group H-2, and Group H-3
- 25 feet in Group H-1, Group H-2, and Group H-3 occupancies
- 100 feet in Group B, Group F, and Group S occupancies when an automatic fire sprinkler system is installed throughout the building
- 100 feet in Group B, Group F, and Group S occupancies when the design occupant load of the space is thirty or less
- 100 feet in Group I-3 occupancies
- 125 feet in Group R-2 occupancies when the building is protected with an automatic sprinkler system

Exits from buildings

After the number of exit access doorways from a room or space is determined, the number of exits from each story must be determined. The process is the same as for rooms and spaces. The only difference is that you are evaluating the entire story. The IBC requires that the occupants in the building all have access to at least two exits in most buildings. In very small buildings, the first and second stories can be provided with only one exit as long as the occupant load is relatively low and the distance a person must travel to get to an exit is relatively short. For the purposes of this discussion, stories with one exit will not be reviewed. Just like rooms and spaces, stories with an occupant load of 501 to 1,000 must be provided with a minimum of three exits and those with more than 1,000 must be provided with four exits. **[Ref. 1021.1]**

Exits are the second part of the means of egress system. When occupants enter an "exit," they are deemed to be protected from the fire. Several different types of exits are provided by the code. The most common type of exit is the exterior door located at ground level of a building (see Figure 10-3). Once you get through this exterior door, you are outside and away from the building and no longer affected by the fire. Other types of exits include interior exit stairways, exit passageways, exterior exit stairways, exterior exit ramps, and horizontal exits. We discuss the interior exit stairways and exit passageways later. Exits are an important part of the means of egress. Remember, they are the protected part of the exit system in a building. An exit must be continuous and lead directly to the exterior of the building. It cannot be used for storage or any use other than providing egress from the building. [Ref. 202]

FIGURE 10-3 Exterior exit doors are considered exits

When an exit terminates at the exterior of a building, it called the *exit discharge*. This is the third part of the means of egress. The exit discharge must be at grade or provide direct access to grade. Once a person is outside the building and in the exit discharge, he or she cannot reenter the building to get to a public way. The occupant must be able to directly access the public way without any obstructions. [Ref. 1027]

Exit paths

Now that we know how much stuff (occupant load) and how many pipes (exits) we have, we now need to figure out how occupants get to the exits in an efficient manner. To provide an effective exit system, the IBC outlines the minimum width of the egress path, the location of the exit access doors or exits, the distance a person has to travel, and the protection of the people within the protected exits. Because people are not always familiar with the design and location of the exit system, the path must be identified and provided with enough light so people can find their way out in a dark and/or smoke-filled atmosphere.

Making it wide enough

The size of the pipes in a plumbing system is based on the amount of stuff going through them. The width of the exit system is based on the same philosophy. The larger the design occupant load, the more exits and the more width needed to efficiently evacuate people from a building. The IBC calculates the width of the means of egress by multiplying the occupant load by a factor to determine the width. The clear width at doors, corridors, and other egress paths other than stairways is calculated by multiplying the occupant load by 0.2 inches. The width of stairs is based on 0.3 inches per occupant. The calculated width must be arranged so that if one of the paths is lost, the total width is not reduced by more than 50 percent. These widths must then be maintained to the public way.

If the building does not have a Group H or I-2 occupancy, the width of the exits can be reduced if the building is provided with an automatic fire sprinkler systems and an emergency voice/alarm communication system. An emergency voice/alarm communication system provides voice instructions as well as evacuation signals in a building during an emergency. The width of the egress system can be reduced to 0.2 inches per occupant for stairways and 0.15 inches per occupant for all other egress components. **[Ref. 1005]**

EXAMPLE

Exit width calculation

Given: 5,000 square foot dining area
Problem: To determine egress width
Solution:

Occupant load:
5,000 ÷ 15 sq. ft. per occupant = 334 occupants
Minimum of two exits required
Exit path width:
334 × 0.2 inch per occupant = 66.8 inches
Two 36-inch-wide doors provide 68 inches of clear width: OK
Stair width
334 × 0.3 inch per occupant = 100.2 inches
Two 50.1-inch stairs required to comply with minimum width
With Sprinklers and Emergency Voice/alarm Communication System
Exit path width = 50.1 inches
The minimum required clear width of a door is 34 inches. Therefore, if two doors complying with this minimum width are provided, they will provide the minimum calculated width as well.
Stair width = 66.8 inches
The minimum required clear width of stairways serving an occupant load of 50 or more occupants is 44 inches. Therefore, if two stairways complying with the 44 inch minimum are provided, they will provide 88 inches of width and comply with the calculated width as well.

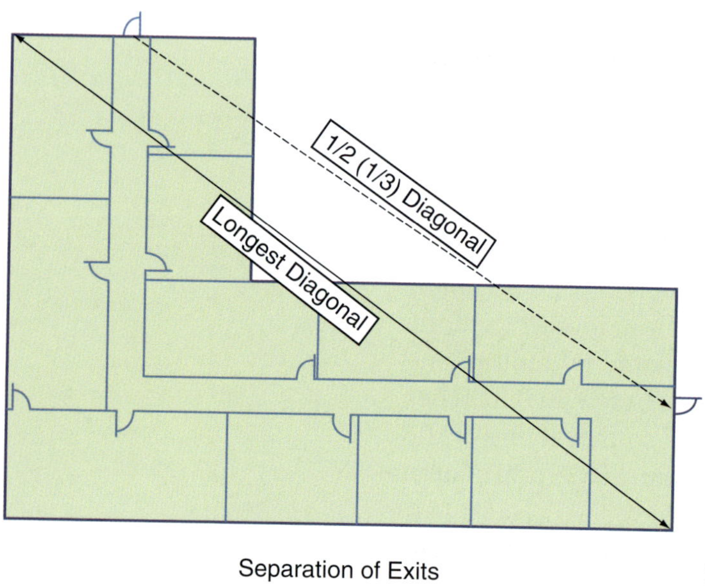

FIGURE 10-4 Separation of exits or exit access door measurement

Horizontal travel

To ensure that occupants have an adequate choice of exits to travel to, the IBC requires that they be separated a minimum distance apart. When two or more exits or exit access doors are required, they must be at least ½ of the longest diagonal of the space that they serve. This applies to each room, space, or story that requires two or more exits. This dimension is measured in a straight line between the two exit doors (see Figure 10-4). When a building is provided with an automatic sprinkler system, the dimension between the exits can be reduced to ⅓ of the longest diagonal. **[Ref. 1015.2]**

In order to provide path of egress within a safe time, the IBC limits the distance a person has to travel in a building to get to a protected

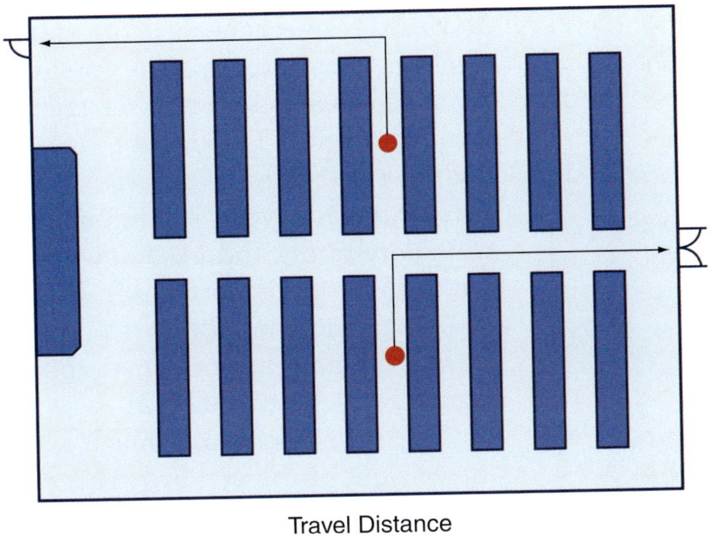

FIGURE 10-5 Travel distance measurement taken along the natural and unobstructed path of egress

TABLE 10-3 Exit access travel distance (Based on IBC Table 1016.1)

Occupancy	Without Sprinkler System (feet)	With Sprinkler System (feet)
A, E, F-1, M, R, S-1	200	250
I-1	Not Permitted	250
B	200	300
F-2, S-2, U	300	400
H-1	Not Permitted	75
H-2	Not Permitted	100
H-3	Not Permitted	150
H-4	Not Permitted	175
H-5	Not Permitted	200
I-2, I-3, I-4	Not Permitted	200

exit. This travel distance is based on the occupancy classification of the building and whether or not an automatic sprinkler system is installed. It is measured along a natural and unobstructed path to simulate the path a person would have to travel to get to an exit (see Figure 10-5). Table 10-3 provides the exit access travel distances for each of the occupancy groups. **[Ref. 1016]**

Vertical travel

In multistory buildings, it is important that people on upper levels have a safe means of evacuation down and out of the building. This is most commonly done by stairs. When stairs are used to get people out of a building, they must be enclosed to protect them from a fire. The fire could be located on a floor that they would need to pass by to exit the building. The fire-resistance rating of the enclosure is based on the number of

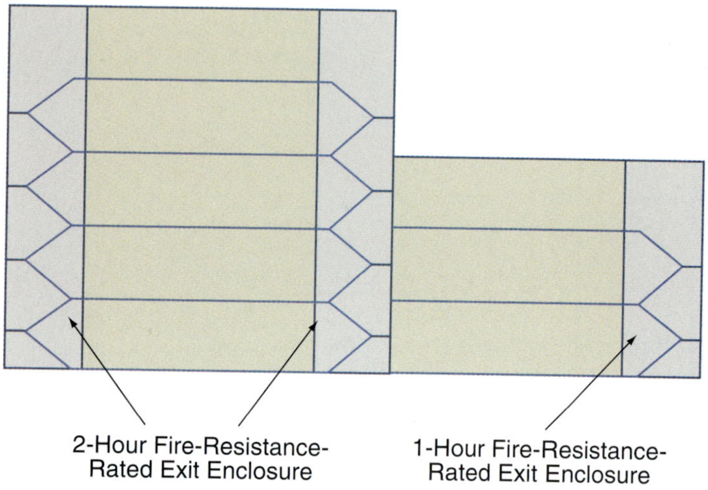

FIGURE 10-6 Exit enclosure fire-resistance-rating requirements

stories the stair serves. If the stair connects two or three stories, the enclosure must be a minimum 1-hour fire-resistance-rated enclosure with 1-hour fire-resistance-rated doors. Enclosures that serve four or more stories must have a minimum 2-hour fire-resistance rating and 1½; hour fire-resistance rated doors. Stair enclosures are considered as "exits" and therefore provide a protected path from the hazards in a building (see Figure 10-6). **[Ref. 1022.2]**

The stair enclosures must extend all of the way out of the building. This can be done either by placing the stair enclosure adjacent to the exterior wall with a door at the bottom directly to the outside of the building or by extending the stair enclosure to the outside via an exit passageway. An *exit passageway* is a horizontal extension of the stair enclosure and is a fire-resistance-rated tunnel that has the same fire-resistance rating as the stair enclosure.

Stair enclosures are not just for people evacuating the building. They are also used for fire department access to the upper levels of the building. The firefighters can also use the stair enclosures to stage their attack of the fire within a fire protected enclosure. In tall buildings, the standpipe connections are required to be installed within the stair enclosures.

Path identification

The location of an exit is not always obvious or known by the people in a building. Therefore, the path to an exit must be identified by exit signs. Exit signs must be visible from any direction that someone may travel along the exit path. They must clearly indicate the direction of the exit path if it is not immediately visible to the people in the building. Exit signs are required whenever a room, space, or building is required to have two exits or exit access doors. The minimum sizes of the sign and the letters are regulated by the IBC (see Figure 10-7). **[Ref. 1011.1]**

FIGURE 10-7 Exit signs over doors identify exit path out of room

Group R-1 occupancies are required to be provided with floor-level exit signs. The bottom of these signs must be located between 10 and 12 inches above the floor level. It must be mounted on either the egress door or within 4 inches of the door frame on the latch side. This gives people a better chance of seeing the exit sign in a smoke filled atmosphere. **[Ref. 1011.2]**

Exit signs must be illuminated from either an internal or external light source. The lighting system must continue to work if the buildingís power is disconnected. This is accomplished either by connecting the exit sign to the emergency power in the building or by signs that have a battery backup system integrated into them. The power source must maintain the illumination for a minimum of 90 minutes. Some internally illuminated exit signs are provided with a self-luminous or photoluminescent system to provide the illumination. Exit signs illuminated from an external source must have an intensity of at least 5 foot-candles of light. **[Ref. 1011.5]**

FIGURE 10-8 Egress illumination with battery standby power

In addition to giving people directions to a safe path of evacuation, the IBC also requires that the path be illuminated at a minimum level. Egress illumination is required throughout the building. The normal lighting system in a building typically provides this illumination. The minimum amount of illumination that is required is 1 foot-candle of light at the floor level. This is just enough light to see any major obstacles that may impede the path of egress. The IBC then requires emergency-powered lighting in case the power goes out. This emergency lighting system is required in areas and buildings that are required to have two exits. It is also required on exterior exit stairways and just outside the exterior exit doors of a building. The emergency system must maintain the power for the same 90-minute period that exit signs do (see Figure 10-8). **[Ref. 1006]**

MEANS OF EGRESS COMPONENTS

As with a plumbing system with valves, elbows, and risers, the exit system has different components. The components that are discussed here are the doors, stairways, and ramps. These are the most common types of components in an exit system and can be located in any of the three parts of the means of egress. The proper design and installation of these components can provide a smooth and safe path for people to get out of a building in an emergency.

Doors

Doors serve as the valves in an exit system. A door can sometimes restrict the flow of people due to the width of the door and the swing of the door. Several different types of doors are used in buildings, the most common being the typical side-swinging door. The IBC requires that all doors in an egress system be this type of door. There are some exceptions for very small areas and special occupancies. Doors used for exiting must have a minimum clear width of 32 inches when the door is open at a 90-degree angle. A typical 36-inch door provides this minimum width. The maximum width of a swinging door leaf is 48 inches. Doors serving an occupant load of 50 or more persons must swing in the direction of travel (see Figure 10-9). **[Ref. 1008.1]**

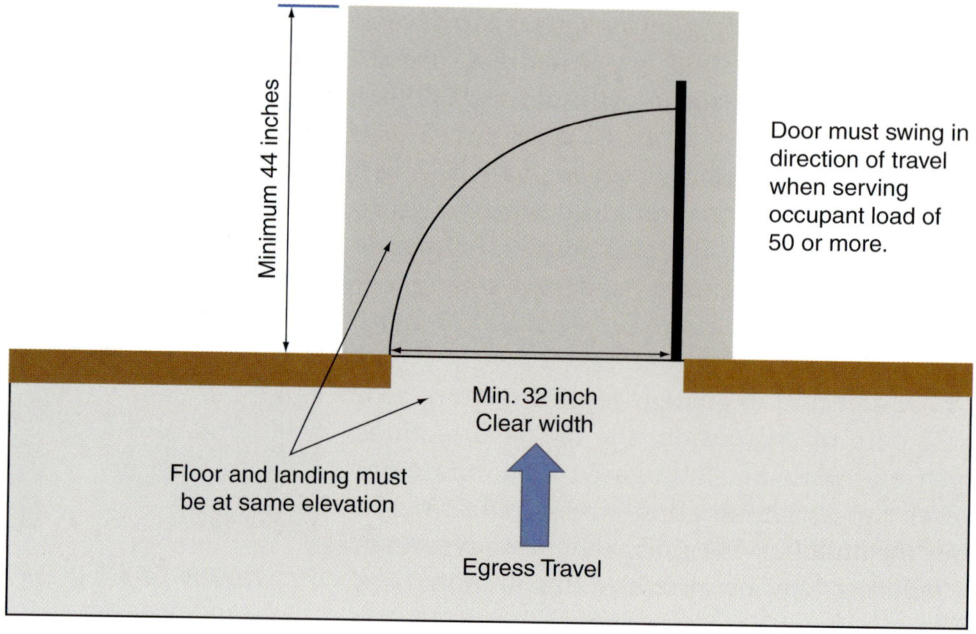

FIGURE 10-9 Egress door requirements

Doors must have a floor or landing on both sides of the opening. When a person walks through a door, he or she typically expects the level on each side of the door to be the same. So, the IBC requires that the floor or landing be at the same elevation on both sides. The landing must be the same width as the door or, if there is a stairway on one side, as wide as the stairway landing. The landing must be a minimum of 44 inches long measured in the direction of travel. To eliminate any potential tripping hazards at a door, thresholds are limited to ½ inch in height. If the threshold exceeds ¼ inch in height, the edges must be beveled at a slope not greater than one vertical in two units horizontal. This bevel makes it easier for people in wheelchairs to get over the threshold. **[Ref. 1008.1.6 and 1008.1.7]**

FIGURE 10-10 Lever-style door handles allow opening of a latch without tight grasping, tight pinching, or twisting of the wrist

The IBC regulates the type of hardware on a door to make it easier for people to quickly get through the door. The most important part of these requirements is that the door must be readily openable from the egress side without the use of a key or special knowledge or effort. When a person comes up to a door, he or she must be able to open it easily. When doors are provided in areas in which people with physical disabilities may be present, the door handle, pull, latch, or lock must be designed so that it can be operated without tight grasping, tight pinching, or twisting of the wrist. The hardware to open the door must be installed between 34 and 48 inches above the finished floor (see Figure 10-10). **[Ref. 1008.1.8]**

Doors in Group H occupancies and doors in Group A and Group E occupancies that serve an occupant load of fifty or more must have panic hardware installed. Panic hardware is activated by pushing on a bar that spans across the face of the door. It does not require any twisting or grabbing to work (see Figure 10-11). Panic hardware is used in these occupancies to provide a faster means of opening the door in an emergency. A person being pushed against the door will unlock and open the door. **[Ref. 1008.1.10]**

FIGURE 10-11 Panic hardware provided in double exit access doors

Stairways

The quickest way for people to evacuate a multistory building is by stairways. The requirement for enclosing stairways was discussed earlier in this chapter. This section discusses the specific requirements for the construction of the stairway itself. Although stairways provide an efficient way to move vertically through a building, they can be very dangerous. One of the main causes of injuries in a building is falls down stairs. It is important that stairways be designed and constructed in accordance with the requirements of the IBC to mitigate these injuries.

Stairways must be wide enough to accommodate the number of people using them. The width must be a minimum of 44 inches measured from wall to wall. To determine the width, the number of people using the stair must be multiplied by a factor of 0.3 inches per occupant. For example, a stair serving three hundred occupants would need to be a minimum of 90 inches in width. If the building is provided with an automatic sprinkler system and an emergency voice/alarm communication system is provided, the required width is calculated at 0.3 inches per occupant. Therefore, a stair that serves 300 occupants, the minimum required width of the stairway would be 60 inches. Stairways also need to be constructed to provide enough headroom that most people will not hit their heads while using them. The IBC requires that stairways have a minimum headroom clearance of 80 inches. This dimension is measured vertically from a line connecting the edge of the stair nosing. The headroom must be maintained continuously along the stairway including the landings. **[Ref. 1009.1]**

To ensure that stairs are safe and uniform in dimension, the IBC specifies the dimensions of a stair riser tread. The maximum height of a riser is 7 inches. The minimum height of a riser is 4 inches. This dimension is measured vertically between the leading edge of the adjacent treads. The minimum length of a stair tread is 11 inches. The tread depth is measured between the vertical planes of the front of the nosing of adjacent treads at a right angle to the treadís leading edge. Figure 10-12 shows these measurements. **[Ref. 1009.47]**

An important part of making a stairway safe is the uniformity of the treads and risers. A small difference in riser height or tread depth can cause someone to fall and injure himself or herself. Therefore, the

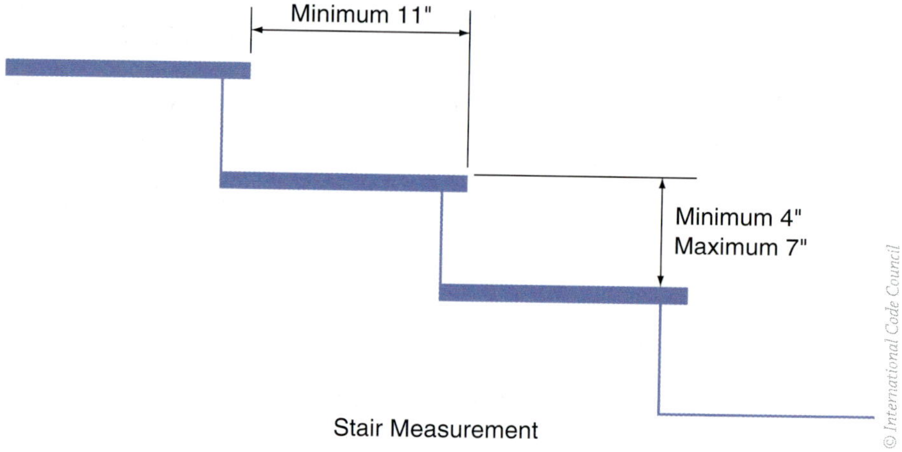

FIGURE 10-12 Stair dimensions

code limits the tolerance between the largest and smallest riser height or tread depth to ⅜ inch. This means that all of the dimensions within a flight of stairs must be within this limit. **[Ref. 1009.47.4]**

All stairways must be provided with landings at the top and at the bottom of a flight of stairs. They must be located so the maximum vertical distance between landings does not exceed 12 feet. The landings must be the same width as the stairs. The lengths must be at least the same as the width. However, when the stairs continue in straight run, the landing need only be 48 inches in length regardless of the width. Doors are permitted to open over a landing, as long as they do not obstruct more than half of the required width of the stair. When a door is fully open, it cannot project more than 7 inches into the landing. To assist people in ascending and descending the stairs, handrails must be provided on both sides of the stair. The details of handrails are discussed in Chapter 12. **[Ref. 1009.8]**

Ramps

A ramp is another way to move to and from different levels of a building. Most ramps are provided as an accessible route for people with physical disabilities. A *ramp* is defined as a sloped walkway that is steeper than one unit vertical in twenty units horizontal (1:20), or a 5 percent slope. Ramp slopes are limited to a maximum slope of 1:12 or 8 percent. Ramps that are not used as a part of the means of egress are limited to a maximum slope of 1:8, or 12.5 percent. The slope across the ramp and landings is limited to a maximum of 1:48 or 2 percent (see Figure 10-13). **[Ref. 1010.2 and 1010.3]**

FIGURE 10-13 Ramp with required handrails on both sides

The minimum required width of a ramp is determined by multiplying the occupant load by 0.2 inches per occupant. As with

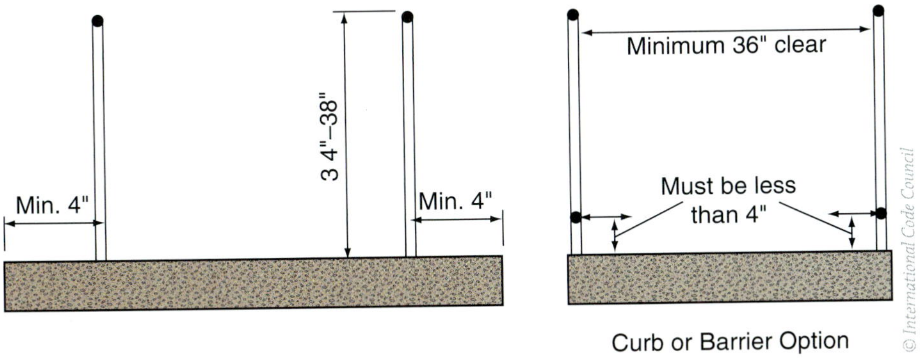

FIGURE 10-14 Ramp edge protection options

stairways, a building that is provided with automatic sprinklers and an emergency voice/alarm communication system, the width of the ramps are based on 0.3 inches per occupant. However, the minimum clear width between the handrails or any obstructions on each side must be 36 inches. The minimum headroom of 80 inches is similar to that required for stairways. Just like stairways, ramps must be provided with landings at the top and bottom. Landings must be as wide as the ramp with a minimum length of 60 inches. The maximum elevation change from landing to landing is 30 inches. Anytime a ramp changes directions at a landing, the landing must be a minimum of 60 inches by 60 inches in size. **[Ref. 1010.5 and 1010.6]**

Ramps that rise more than 6 inches must be provided with a handrail on each side. The handrail height and graspability requirements are the same as stairways. The extensions are a little different in that they need to extend 12 inches minimum beyond the top and bottom of the ramp runs. Because ramps are commonly used by people in wheelchairs, the IBC requires edge protection on each side of ramps and their landings. Edge protection can be done either by a curb, rail, wall, or barrier or by extending the surface of the ramp a minimum of 12 inches beyond the inside face of a handrail. When a curb is used, it must be a minimum of 4 inches in height. Rails and barriers must be constructed so a 4-inch sphere cannot pass through an opening between the ramp and barrier/rail (see Figure 10-14). **[Ref. 1010.10]**

If the means of egress is designed and constructed in accordance with Chapter 10 of the IBC, the occupants of the building will have a safe means of evacuating a building in an emergency. The three parts of the means of egress—exit access, exit, and exit discharge—must be understood to correctly use the code. Attention to the detailed requirements for stairways, doors, and other components is also important to the proper design of an area.

CHAPTER 11

Getting People In

© International Code Council

The *International Building Code* does not have any specific requirement that says someone must be able to enter a building. Such a requirement is really not needed, because every building has an entrance. However, if a person has a physical disability, he or she must be provided with the same level of access as a person without such disabilities. The previous chapter discussed what happens after people are in the building and they need to get out in the event of an emergency. This chapter will discuss the requirements in the code that provide a person with physical disabilities access to services just like anyone else.

FEDERAL REGULATIONS

In 1990, President George H. W. Bush signed the American with Disabilities Act (ADA). It was a civil rights legislation to provide a clear and comprehensive national mandate for the elimination of discrimination against individuals with disabilities. It also provided a clear, strong, consistent, enforceable standard addressing discrimination against individuals with disabilities. The ADA is enforced by the federal Department of Justice through civil rights litigation. The ADA references the American with Disabilities Act Accessibility Guidelines (ADAAG) for buildings and facilities. This document provides regulations that outline the building-related requirements. They are applied during the design, construction, and alterations of buildings regulated by the ADA.

Prior to the ADA, the federal government passed the Federal Fair Housing Act (FFHA) to provide accessible living facilities for people with disabilities. The act applies to buildings consisting of four or more dwelling units if the building has one or more elevators; or any ground floor dwelling units in buildings consisting of four or more units. The intent is to create equal housing opportunities for all persons living in the country. The FFHA is enforced by the U.S. Department of Housing and Urban Development (HUD). It also has design guidelines that provide the minimum requirements with which the design and construction must comply.

In addition to these federal standards, the IBC also has accessibility requirements for buildings. The IBC requirements are separate from the federal standards and are enforced by the local jurisdiction that adopts the code. Local jurisdictions do not have the authority to enforce either the ADAAG or the FFHA. However, the regulations in the IBC and associated ICC Standard for Accessible and Usable Buildings and Facilities (A117.1) are very similar to the federal standards. In fact, the IBC and A117.1 have been accepted by HUD as "safe harbor" under the FFHA. This means that if a building is designed to the IBC and A117.1, the design is deemed to comply with the FFHA. Attempts to get the same approval under the ADA are ongoing.

ACCESSIBILITY

The IBC contains *scoping requirements* for accessibility. Scoping requirements tell the user when a building must be accessible and to what extent. The A117.1 standard provides the requirements on how to make the building and facility accessible. The two documents must be used together to accomplish the intent of making buildings accessible to the physically disabled.

The scoping requirements of the IBC start out by requiring all sites, buildings, structures, facilities, elements, and spaces, temporary or permanent, to be accessible to people with physical disabilities. At first glance, this may seem to be a very extreme requirement. However, the code lists several situations for which accessibility is not required.

Detached one- and two-family dwellings and their accessory structures do not need to be accessible. Construction sites are not required to be accessible for obvious reasons. Spaces that contain equipment that are frequented by maintenance personnel for repair and monitoring do not have to be accessible. Walk-in coolers and freezers are not required to be accessible either. **[Ref. 1103]**

Accessible path

The IBC requires an accessible path to the building and through the building. This is accomplished by providing an "accessible route." The accessible route is a continuous, unobstructed path that complies with the IBC and A117.1. It can be a walkway, doors and doorways, ramps, curb ramps, elevators, and platform lifts. All of these components of the accessible route can be used by a person in a wheelchair or by a person with some other form of physical disability. To provide equality for disabled persons, the accessible route must coincide with or be located in the same area as the general circulation path. If the circulation path is on the inside of the building, the accessible route must also be inside the building. The disabled person cannot be required to go outside the building to get to a space where an unimpared person could get there by just going through the building. **[Ref. 1104.5]**

Into the building

An accessible path must be provided from points where people may arrive at the property at which they want to visit or do business. Therefore, the IBC requires an accessible route from public transportation stops (bus stops), accessible parking spaces, accessible passenger loading zones, and public streets or sidewalks to the accessible building entrance (see Figure 11-1). **[Ref. 1104.1]**

Once a person is on a site, he or she also must be able to get from one building to another building if there are multiple structures on a property. Therefore, an accessible route must connect any accessible buildings, facilities, elements, or spaces on the same site. For example, in a multibuilding apartment complex, the clubhouse and recreational facilities are required to be accessible. Therefore, an accessible route must be provided from the accessible dwelling units to the clubhouse and recreational facilities. **[Ref. 1104.2]**

FIGURE 11-1 Accessible route from public street to building entrance

At least 60 percent of the public entrances into a building must be accessible. This means that if two public entrances are provided, both must be accessible. Accordingly, if there is only one entrance, it must also be accessible. In buildings with multiple tenant spaces, dwelling units, and sleeping units, at least one accessible entrance must be provided to each tenant space or unit.

Through the building

After a person with a disability gets in to the building, he or she needs to get through the building to conduct whatever business he or she intends to do. The IBC requires that an accessible route be provided to each portion of the building from the accessible entrance. Multistory buildings and facilities are required to have an accessible route to each accessible level of the building. If a story is located above or below the accessible level and the story is 3,000 square feet or less in floor area, an accessible route may not be required. This exception is not applicable if there are five or more Group M tenants in the building; if a level contains health-care provider offices; or for passenger transportation facilities and airports. There are two ways to provide an accessible route between two different levels in a building: ramps and/or elevators.

Ramps

The first question that needs to be answered is, When is a ramp a ramp? A ramp is defined in the IBC as a walking surface that has a running slope steeper than one unit vertical in 20 units horizontal. That is translated into a 5 percent slope. If the slope is less than 5 percent, it is considered a walking surface and not a ramp. (The construction of ramps was discussed in the previous chapter.) Ramps are a very common way of providing an accessible route to a building and throughout the building. To be used as accessible routes, they cannot exceed a 1:12 slope.

Elevators

In multistory buildings, elevators are the most common and most efficient way to provide an accessible route to upper or lower floors. Elevator cars must be large enough to get a person in a wheelchair and one additional person in the elevator. The controls must be placed so that a person in a wheelchair can reach them to operate the elevator. Elevators are typically manufactured to comply with the accessibility provisions (see Figure 11-2). In certain conditions, a platform or wheelchair lift can be used as part of accessible route in a multi-story building. **[Ref. ICC A117.1 Sect. 407.4.1]**

FIGURE 11-2 Elevators are the primary vertical accessible route. Notice trash can obstructing fire door

Usable spaces

When a person with physical disabilities is in a building, he or she must be able to use the facilities inside. Persons with physical disabilities should be able to move through a building and enjoy the services the same as anyone else. The IBC requires that specific facilities and areas of a building be accessible. We discuss a few of those locations, including toilet facilities, kitchens, and assembly seating areas.

Toilet and bathing facilities

The IBC requires that toilet rooms and bathing rooms be accessible. All such facilities must be accessible, even though they may not be located on an accessible level. In addition, the *only* toilet rooms or bathing rooms cannot be located on a level that is not accessible. At least one of each type of fixture, element, control, and dispenser in the toilet room or bathing room must be accessible.

Assembly and mercantile occupancies have a higher probability of having a disabled person as a visitor. In some cases, these people have another person to assist them. Therefore, the code requires these occupancies be provided with an accessible family or assisted-use toilet room if there are a total of six or more water closets required in the men's and women's restroom combined. This prevents the concern of people having to enter a restroom of people of the opposite sex. These toilet rooms are separate from the separate-sex toilet rooms but can be counted toward the total required plumbing fixtures in a building.

Family or assisted-use toilets must be located on an accessible route. They must be located so that they are no more than 500 feet from separate-sex toilet rooms and no more than one story above or below them. They are limited to only one water closet and one lavatory. Family or assisted-use bathing rooms are also limited to one water closet, one lavatory, and one shower or bathtub. Doors to family or assisted-use toilet and bathing rooms must be securable from the inside of the room to provide privacy.

Water closets

As noted previously, at least one of each fixture in a toilet room or bathing room must be accessible. Therefore, if a toilet room has multiple water closets, at least one of the water closets must be wheelchair accessible and comply with the requirements of ICC A117.1. To make the toilet room usable to a person in a wheelchair, the area must be large enough to provide room for that person to move freely into and through the room. The first requirement in Standard A117.1 is the turning space. The standard requires a minimum 60-inch-diameter turning space. As an alternate, a T-shaped space can be provided (see Figure 11-3). These turning spaces provide the needed area to maneuver a wheelchair in the room. The turning space is a common requirement for accessible spaces.

Toilet compartments must have a minimum dimension of 60 inches wide and either 56 inches deep if a wall-hung water closet is installed or 59 inches deep if a floor-mounted water closet is installed. This allows a person in a wheelchair to enter the compartment, back up next to the water closet, and make a side transfer to the fixture (see Figure 11-4). Single water closets located in a single toilet room must have a clear space of 60 inches wide and 56 inches deep. No other fixtures like lavatories are permitted to be located within this clear space. The door serving the toilet compartment or room must be wide enough

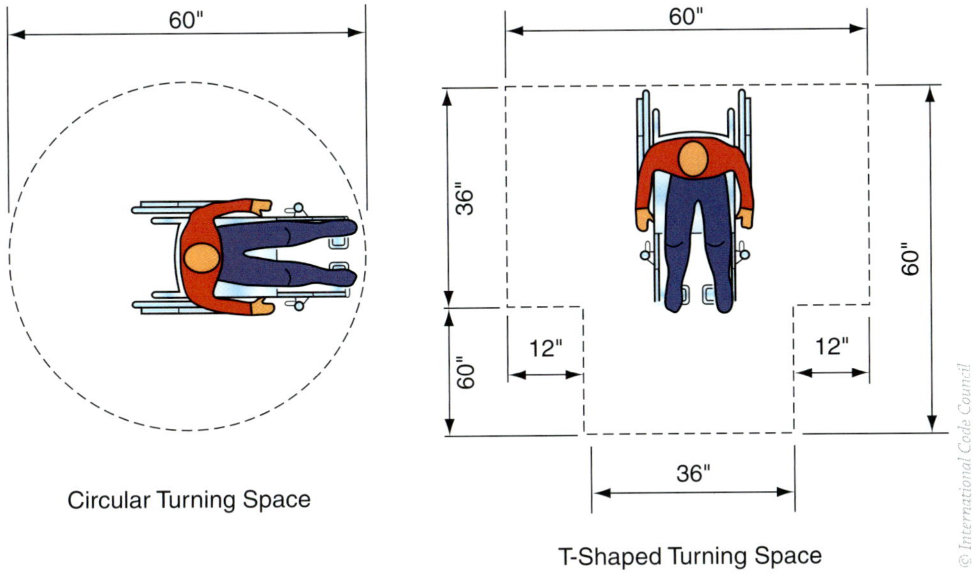

FIGURE 11-3 Required wheelchair turning space

so that a wheelchair can get through. There must also be enough room in front of the door to get into the compartment.

Within the compartment, grab bars must be provided on the rear and side walls. The rear grab bar extends behind the water closet to provide a handle for a person with a disability to lift and move themselves on and off of the water closet. It must be located between 33 and 36 inches above the finished floor. Two side grab bars are required, one horizontal and one vertical. The vertical bar provides a grabbing point to pull oneself up from the water closet (see Figure 11-5). **[Ref. 604.5 (A117.1)]**

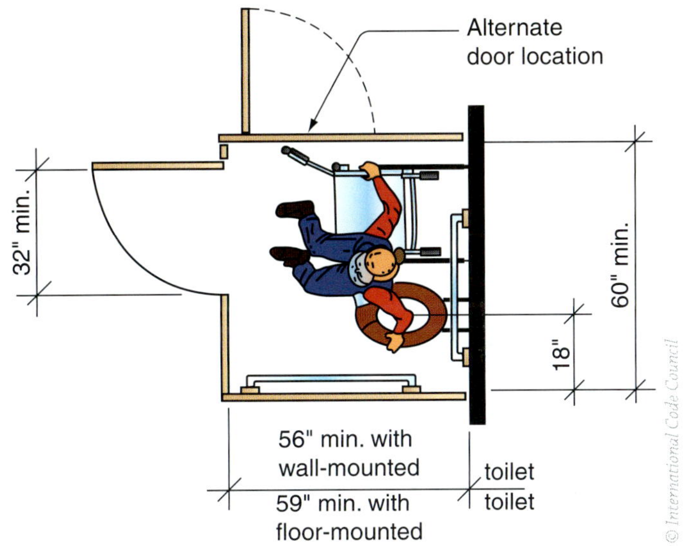

FIGURE 11-4 Wheelchair transfer

Water closets are permitted to be either wall hung or floor mounted. They must be located with a wall at the rear and to one side. The water closet must be located a specific dimension from the side of the partition or wall so the person does not have to reach too far to use the grab bars. The flush controls must be hand operated or automatic. However, if it is a hand control, it must be located on the open side of the water closet. As a person in a wheelchair maneuvers inside a toilet compartment, his or her feet extend underneath the partition. Therefore, the ICC standard requires a minimum toe clearance. **[Ref. 604.2, 604.4, and 604.6 (A117.1)]**

Lavatories and sinks

At least 5 percent but not less than one of the lavatories in a toilet facility must be accessible. There must be enough room to get to and use the lavatory. Therefore, Standard A117.1 requires that a clear floor space be provided in front of the sink to allow the person in a wheelchair the room to use the sink. In order for a person in a wheelchair to get up to the sink

FIGURE 11-5 Grab bars provided on side and rear of water closets

and use it, he or she must extend his or her legs under the lavatory. Therefore, the standard requires a minimum knee and toe clearance under the fixture (see Figure 11-6). **[Ref. 606.2, 305.3, and 306 (A117.10)]**

The front of the lavatory must be a maximum of 34 inches above the floor. The faucet must be operable with one hand and must not require tight grasping, pinching, or twisting of the wrist; or it must be a hand-operated metering faucet that remains open for a minimum of 10 seconds. Any water supply and drain-pipes below the lavatory must be insulated or protected so that persons in a wheelchair without feeling in their legs do not burn themselves. Sharp or abrasive surfaces are also prohibited under lavatories for the same reason. **[Ref. 606 (A117.1)]**

Kitchens

When kitchens are provided in a building, they must be accessible to people who are physically disabled. A kitchen includes kitchenettes and wet bars. This section discusses kitchens that are not part of a residential unit. Residential kitchens are required to meet different standards than kitchens in a nonresidential occupancy. Commercial kitchens serving restaurants are considered employee work areas and are only required to be on an accessible route and be designed so a disabled person can approach, enter, and exit the area. The kitchen equipment is not required to be accessible. Therefore, this section only discusses kitchens located in break rooms and similar areas.

Two types of kitchens are covered by Standard A117.1: pass-through and U-shaped. A pass-through kitchen is just like it sounds; a person can go completely through the kitchen. Typically, the counters, appliances, and cabinets are on two opposing sides. The minimum clearance between the counters, cabinets, and appliances is 40 inches. This allows a person in a wheelchair to use the kitchen in a side-approach manner. A U-shaped kitchen must have a minimum clearance of 60 inches between all of the counters, appliances, and cabinets. This allows someone in a wheelchair to enter and turn around in the kitchen (see Figure 11-7). **[Ref. 804.2 (A117.1)]**

The countertops must be between 28 and 34 inches above the finished floor. If the kitchen does not include a cooktop or range, the kitchen sink can be used by a person who is in a wheelchair set parallel to the counter. However, if there is a cook-top or range, a forward approach must be provided to the sink, similar to that discussed regarding lavatories. The required

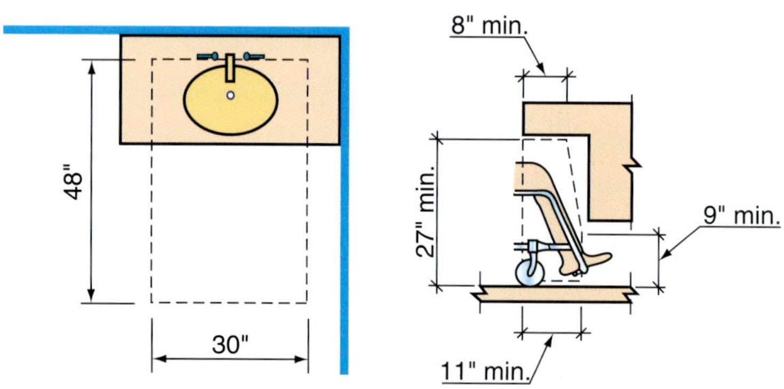

Clear Floor Space at Lavatory

Knee and Toe Clearance

FIGURE 11-6 Required clearances for lavatories

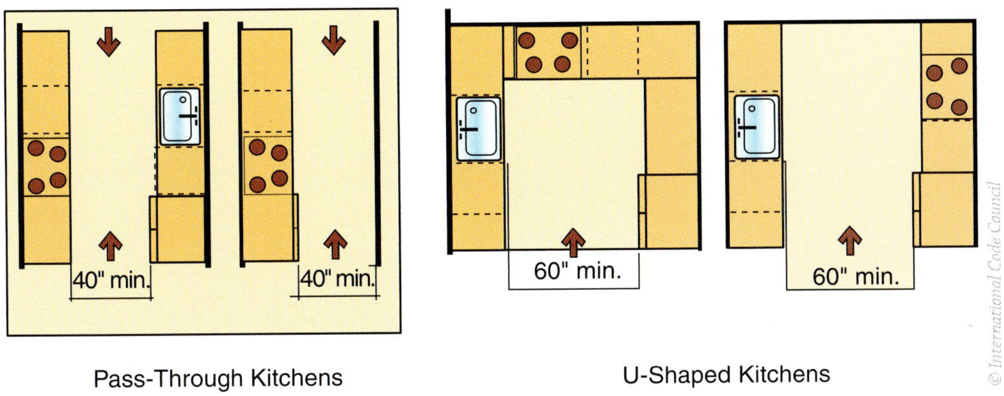

FIGURE 11-7 Kitchen options

knee and toe spaces noted for lavatories must be provided. The faucets must also be operable with one hand and must not require tight grasping, pinching, or twisting of the wrist; or they must be hand-operated metering faucets that remain open for a minimum of 10 seconds. The piping below the kitchen sink must also be insulated. **[Ref. 804 (A117.1)]**

If appliances are provided in a kitchen, they must also be usable by a person with a physical disability. Two requirements need to be addressed regarding the accessibility of appliances. The first is the clear floor space in front of the appliance. The second is the location of the operable parts of the appliance. The clear floor spaces are dependent on the type of appliance. A dishwasher must have a clear floor space adjacent to the dishwasher. A range or cooktop is permitted to have either a forward or parallel approach. If a forward approach is provided, the required knee and toe space must be provided. Ovens must have a clear floor space adjacent to the door. The clear floor space in front of a refrigerator/freezer must be a parallel approach. All of the operable parts of the appliances must be within the required reach range.

Assembly seating areas

To provide equal accessibility in assembly areas, the IBC requires accessible seating areas. Dining areas are generally required to have the total areas allotted for seating be accessible. However, there are two exceptions to this requirement. Levels that are not on an accessible route and contain less than 25 percent of the total area of the dining area and have the same services available on an accessible route are not required to be accessible. In sport facilities, at least 25 percent of the dining area in tiered dining areas must be on accessible routes. At least 5 percent but not less than one of the dining surfaces for seating and standing spaces must be accessible. The accessible dining surfaces must be distributed throughout the facility. They must also be located on a level that is on an accessible route. **[Ref. 1108.2.9]**

In assembly occupancies with fixed seating, the requirements are more specific. Wheelchair spaces must be provided in theaters, bleachers, grandstands, stadiums, arenas, and other similar fixed seating assembly

TABLE 11-1 Accessible wheelchair spaces (Based on IBC Table 1108.2.2.1)

Capacity of Seating in Assembly Areas	Minimum Required Number of Wheelchair Spaces
4–25	1
26–50	2
51–100	4
101–300	5
301–500	6
501–5,000	6, plus 1 for each 150, or fraction thereof, between 501 through 5,000
5,001 and over	36 plus 1 for each 200, or fraction thereof, over 5,000

FIGURE 11-8 Accessible seating spaces next to companion seats. Note additional row of accessible seating at lower level

areas. The spaces must be provided in the numbers outlined in Table 11-1. When luxury boxes, club boxes, and suites are provided, they must also have wheelchair spaces in accordance with this table. Other boxes must be provided with wheelchair spaces in not less than 20 percent of all of the boxes. Even team and player seating areas must be provided with at least one space in each team or player seating area (Figure 11-8). **[Ref. 1108.2.2]**

The wheelchair spaces must be dispersed throughout a multilevel assembly seating area. They must be provided on the main floor level and one of each two additional floor or mezzanine levels. As previously noted, they must also be provided in the luxury boxes, club boxes, and suites if the facility has them. At least 5 percent but no less than one of the total number of aisle seats provided in a facility must be designated accessible aisle seats. These seats must be located on an accessible route. In addition to the wheelchair spaces, at least one companion seat must be provided for each wheelchair space.

Wheelchair spaces are required to be a minimum of 36 inches wide. If two spaces are provided next to each other, they only need to be 33 inches wide. If the space can be accessed from the front or rear, it must be at least 48 inches deep. However, if it can only be accessed from the side, it must be at least 60 inches deep (Figure 11-9). The companion seat must be provided beside the wheelchair space. It must be comparable in size and quality. It must also be on the same elevation as the wheelchair space floor surface. **[Ref. 802 (A117.1)]**

One of the most difficult requirements regarding wheelchair spaces is the "line-of-sight" requirement. The standard requires that the space be located in such a manner that the person in a wheelchair can see over the heads of the people in front of him or her. In assembly areas where people are expected to be seated during the performance, such as a movie theater, the space must be elevated so that the person in a wheelchair can either see over the heads of people in the first row in front of them or over the shoulders and between the heads of the people seated in the first row in front of them. If the people in the assembly area are expected to stand during the event, such as in an arena or stadium, the wheelchair space must be placed so that the person in the wheelchair can see the event over the head of the person two rows in front of him or her.

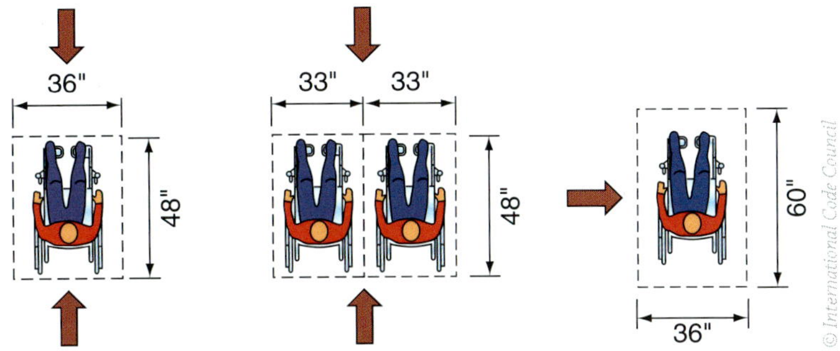

FIGURE 11-9 Wheelchair space dimensions

In addition to the wheelchair spaces, the code also requires assembly occupancies to provide assistance for people who are hearing impaired. If audible communication is an integral part of the assembly space and there is audio amplification, assisted-listening systems must be provided. Assisted-listening systems are typically small radio receivers with headphones. Hearing-impaired patrons can request the receivers so they can hear the event better.

RESCUE FOR PEOPLE WITH PHYSICAL DISABILITIES

This chapter has been dedicated to explaining how buildings must be designed to provide access for people with physical disabilities. This section discusses how to get those people out of a building in an emergency. Obviously, a person in a wheelchair cannot use the same path of egress that an able-bodied person can use in a multistory building. Therefore, the IBC requires that all floors that are accessible be provided with an "accessible means of egress." This term is a little misleading in that for a multistory building, it is more of an accessible means of rescue. The general intent is that if persons with physical disabilities cannot egress the building on their own, the building will be constructed in such a way that these persons are protected until the fire department can reach them and rescue them from the building.

The code requires at least one accessible means of egress from each accessible space. However, when two or more means of egress are required from a space, a minimum of two accessible means of egress must be provided. They must be continuous to the public way and consist of one or more components that make up the accessible path or means of rescue. In buildings that are generally five or more stories in height, an elevator must serve as one of the required accessible means of egress. **[Ref. 1007]**

The accessible means of egress from a single-story building can be accomplished with walking surfaces and ramps that provide an egress path for a person with physical disabilities. In some cases, the difference in elevation from the landing at an exit door and the adjacent grade is so

You Should Know

Accessible Means of Egress Components

- Accessible route
- Interior exit stairways
- Exterior exit stairways
- Elevators
- Platform lifts when used for access
- Horizontal exits
- Ramps
- Areas of refuge ●

large that a ramp is not a viable option. If this is the case and a stairway is provided, the code requires that an area of rescue assistance be provided. This area is an enlarged landing open to the outside. It is designed so a person in a wheelchair can wait to be rescued by the fire department. The landing must be able to accommodate a wheelchair as well as maintain the path of egress for other people in the building. The enlarged landing is protected from the building by a 1-hour fire-resistance-rated exterior wall that extends a minimum of 10 feet on each side of the area. The openings in this separation are required to have a 45-minute fire-resistance rating. The stairway serving the exterior area of rescue assistance must have a minimum clear width of 48 inches between the handrails. This provides the needed width so three people can carry a person and his or her wheelchair down the stairs (Figure 11-10). **[Ref. 1007.7]**

The two most common components used to provide an accessible means of egress from a multistory building are interior exit stairways and elevators. When an interior exit stairway is used as the accessible means of egress, it must meet additional requirements over and above the requirements outlined in Chapter 10. Obviously, a person in a wheelchair cannot use the stairway safely to get out of the building. Therefore, the intent of the code is to provide a safe place inside the stair enclosure for the person to wait and be rescued. This area is called an *area of refuge*. When a stairway is used as an accessible means of egress, an area of refuge must be incorporated into the stair enclosure or the stairway must be accessed through an area of refuge. The details of the area of refuge are discussed later in this section. In addition to the area of refuge, the clear width between the two handrails in the stairway must be a minimum of 48 inches so the person in the wheelchair can be carried down the stairs. If the building is provided with an automatic fire sprinkler system, however, the area of refuge and the minimum 48-inch clear width between handrails are not required. The sprinklers are designed to suppress the fire and provide enough protection for the disabled person to be rescued. **[Ref. 1007.3]**

When an elevator is used as the accessible means of egress, it must also comply with additional requirements. The elevator must be accessed

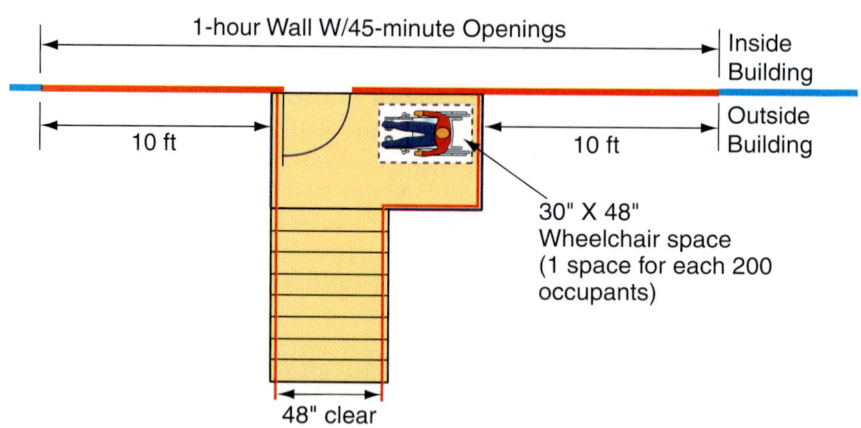

FIGURE 11-10 Areas of assisted rescue on outside of building

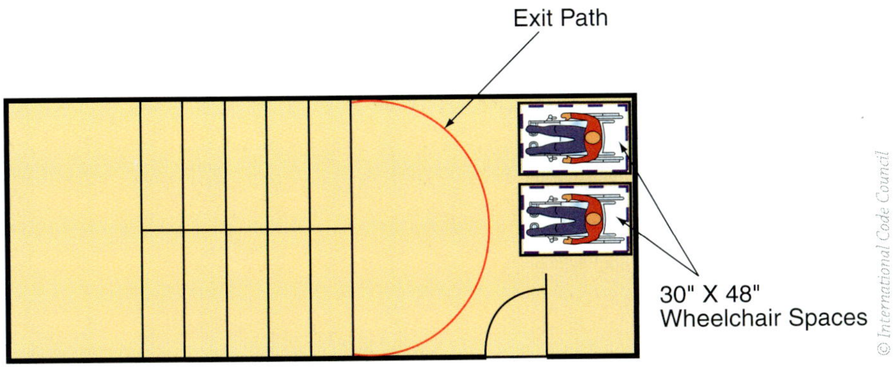

FIGURE 11-11 Area of refuge located in interior stairway enclosure

from an area of refuge or horizontal exit unless the building is provided with an automatic fire sprinkler system. It must be provided with standby power. The standby power must transfer within 60 seconds after the failure of normal power. When multiple elevators are provided, they much each return to a designated level and open the doors. When they have all reached the designated floor, at least one elevator must remain operable from the standby source. **[Ref. 1007.4]**

An area of refuge is designed to provide a safe place for a disabled person to wait to be rescued. Areas of refuge are typically located in enlarged landings of interior exit stairs or are located in elevator lobbies when the elevators are used as the accessible means of egress. Each area of refuge must be sized to accommodate a wheelchair space of 30 inches by 48 inches. At least one space must be provided for every 200 occupants served by the area of refuge and must be located so it does not obstruct the exit path within the stair enclosure (see Figure 11-11). When the area of refuge is not located within a stair enclosure, it must be separated from the rest of the building by a 1-hour smoke barrier to minimize the intrusion of smoke. **[Ref. 1007.6]**

To notify rescuers that a person is in need of help, the IBC requires that a two-way communication system be provided at the elevator landing of each floor of a multistory building or within an area of refuge if provided. The system must provide communication between the elevator landing or area of refuge and a central control point or fire command center if provided. If a central control point is not provided or it is not attended on a continuous basis, the system can be designed to dial 911 directly. The system must also include both audible and visible signals. Directions on how to use the system must be provided so a person in need of rescue can summon assistance. **[Ref. 1007.8]**

CHAPTER

12 Building Safety

© International Code Council

Another part of life safety protection provided by the building code is building safety. Building safety is protection for the occupants of the building from hazards other than fire and emergencies. Buildings inherently have dangerous areas where people can fall and injure themselves. This chapter discusses these areas and how the code provides requirements to protect people from such injuries.

PROTECTION FROM FALLING

As people move through a building, they may be walking along an elevated level from which a fall over the edge would hurt them. Stairways are also dangerous areas where people can fall and get injured. Recent changes to the code have identified that, in addition, children are injured from falling out windows. The IBC provides requirements to mitigate the potential problems with these areas of a building.

Guards

Whenever there is an open-sided walking surface located more than 30 inches above the floor or grade below, a guard must be provided to prevent someone from falling to the level below. The 30-inch measurement is taken at the lowest point within 36 inches horizontally from the edge of the open side of the walking surface. The top of the guard is required to be a minimum of 42 inches above the adjacent walking surface, or the leading edge of stair treads (see Figure 12-1). When mechanical equipment or roof hatch openings are located within 10 feet of the edge of the roof, a guard must be provided to protect service personnel from falling off of the roof. **[Ref. 1013]**

A guard must be constructed to withstand the weight of a person leaning against it. It must be designed to resist a 200-pound concentrated load against the top of the rail applied in any direction. It must also resist a uniform load of 50 pounds per linear foot. **[Ref. 1607.7.1]** Any openings in a guard must be designed so that a 4-inch-diameter sphere cannot pass through the opening. The 4-inch dimension is based on the average size of an infant's head. If the intermediate rails are spaced properly, the chance of an infant getting through the opening and falling through is mitigated (see Figure 12-2). Because it is likely that infants will not be present in areas that are not open to the public or those areas located on unoccupied roofs, the openings are permitted to be placed so that a 21-inch sphere cannot be passed through them.

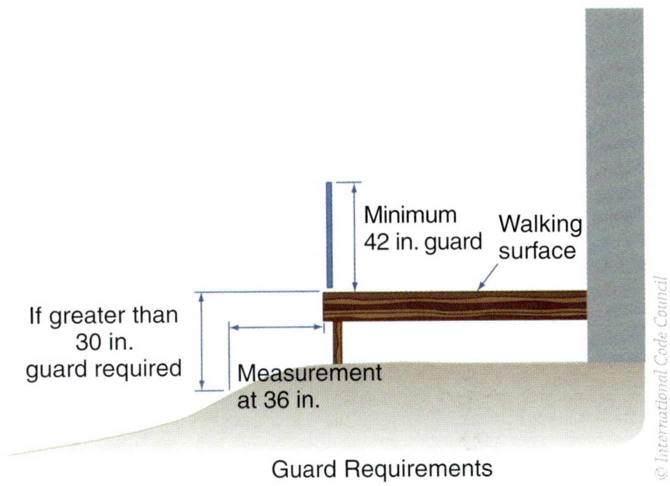

FIGURE 12-1 Guard measurement requirements

FIGURE 12-2 Guard on elevated stairway with the required spacing of balusters

Stairways

Stairways are inherently dangerous when it comes to falls. The U.S. Consumer Product Safety Commission identified over 33,000 reported cases of people falling down stairs in 2008. A large number of those injured were older people. To reduce the number of injuries caused by people falling as they travel on the stairs, the code requires both guards and handrails. The guards are required anytime the height of the stair walking surface is more than 30 inches above the floor or grade below. They prevent someone from falling over the side of the stairway. Handrails are required on both sides of stairs regardless of height with a few exceptions. The handrails are used to assist a weaker person in ascending the stairs and to help someone who falls and tries to catch himself or herself on the handrail. **[Ref. 1009.12]**

Guards adjacent to stairways must be 42 inches in height measured above a line connecting the leading edge of the treads. The openings in a guard along a stairway have the same 4-inch limitation as guards along a walking surface (see Figure 12-3). However, there is an exception for the triangle that is formed by the tread, riser, and bottom rail of the guard. At this location, the opening is limited to a size at which a 6-inch sphere cannot be passed through it.

In addition to the guard along the side of a stair, a handrail must be provided. The top of the handrail must be located between 34 and 38 inches above the tread nosing. It must be graspable so that people can either pull themselves up the stairs or, if they begin to fall down the stairs, they can catch themselves. There are two types of handrails: Type I and Type II. A Type I handrail is the most common and can either be circular or noncircular. Circular handrails must have an outside diameter between 1¼ and 2 inches. Noncircular handrails must have a perimeter dimension of at least 4 inches and not greater than 6¼ inches and a maximum cross-section dimension of 2¼ inches. A Type II handrail is a railing that exceeds the limits of a Type I handrail and is permitted to be used only within dwelling units. This handrail must meet specific dimensions and have finger recesses on each side of the railing to provide a graspable surface. **[Ref. 1012.3]**

Handrails must be continuous along the stairway. They cannot be interrupted by newel posts or other obstructions. When people walk

FIGURE 12-3 Circular handrail along both sides of stairway

down the stairs, they should be able to maintain their grip on the handrail as they use the stairs. When stairs reach a landing, the handrail must either continue to the next flight of stairs or terminate in a specific manner. Handrail extensions must be provided when the handrail is not continuous. The handrail must extend horizontally at least 12 inches beyond the top riser and continue to slope for a depth of tread beyond the bottom riser. The handrail must then return to a wall, guard, or the walking surface so that people do not get their sleeves caught on the end of the handrail. Handrails must also be held away from a wall or other surface a minimum of 1½ inches. The surfaces adjacent to handrails must be free of any sharp or abrasive elements to eliminate the possibility of people cutting or scratching themselves as they use the railing. **[Ref. 1012.4]**

Ramps

Ramps are typically used by persons who may have a disability. Therefore, ramps are also required to have handrails installed on each side. The requirements for ramp handrails are similar to those for stairways. If the ramp is constructed with the walking surface more than 30 inches above a surface below, a complying guard is required to be installed. Ramps must also be provided with edge protection to prevent someone in a wheelchair from rolling off the edge of the ramp.

Edge protection can be provided in two different ways. The first option is to provide a curb, rail, wall, or barrier along the side of the ramp. If a curb or rail is used, it must be a minimum of 4 inches in height. If a barrier is used, it must be constructed so that a 4-inch-diameter sphere can not be passed through the barrier anywhere within the 4-inch height. The second option is to extend the ramp surface a minimum of 12 inches beyond the inside face of both handrails. **[Ref. 1010.10]**

Window-sill height

Reports of children falling through window openings have increased over the last few years. Therefore, the IBC includes requirements to protect children from falling out of windows. Such openable window protection is only required in Group R-2 and Group R-3 occupancies. Any operable window in an exterior wall that has a sill that is located more than 6 feet above grade or other surface below must be provided with fall protection. This is typically accomplished by locating the height of the window sill a minimum distance above the floor surface. The lowest part of the clear opening of the window must be a minimum of 36 inches above the finished floor (see Figure 12-4). For windows that are located less than 36 inches above the floor, the glazing must be fixed or restricted from opening to a maximum dimension that will allow a 4-inch sphere to

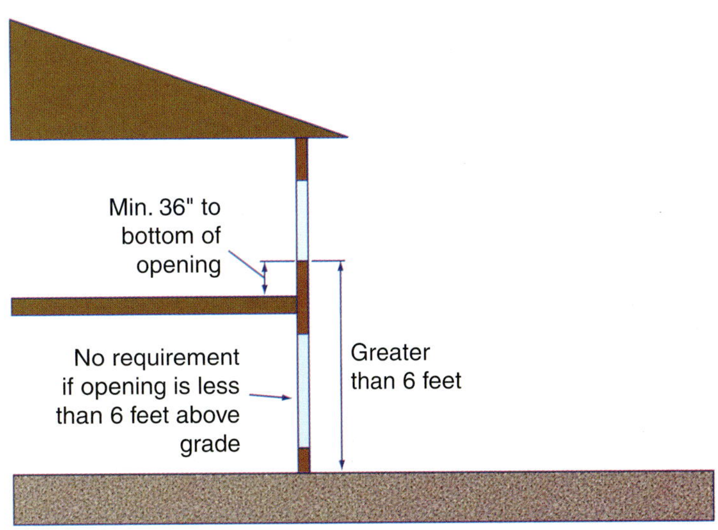

FIGURE 12-4 Window fall protection requirements

pass through the opening. Window fall prevention devices that comply with ASTM F 2090, or windows that are provided with window openings control devices that comply with Section 1013.8.1, can also be used in lieu of the minimum sill height and opening restrictions. **[Ref. 1013.8]**

SAFETY GLAZING

When regular glass breaks, it creates large, sharp, and dangerous shards that can injure whoever comes in contact with them. A person falling through such glazing can be hurt seriously. Therefore, the IBC outlines requirements for the location of safety glazing to mitigate the most common ways people can get hurt when impacting glass. Safety glazing requirements were developed by the U.S. Consumer Product Safety Commission (CPSC) and have been incorporated into the code. To be considered safety glazing, glass must meet CPSC 16 CFR Part 1201 Safety Standard for Architectural Glazing Material. Glazing that is not located in (1) doors or (2) enclosures for hot tubs, whirlpools, saunas, steam rooms, bathtubs, and showers is also permitted to be tested in accordance with American National Standards Institute (ANSI) Z97.1. These standards test the glass for strength and size of the pieces created when the glass is broken. **[Ref. 1406.2]**

Two major types of glazing meet the standards for safety glazing: laminated glass and tempered glass. Laminated glass is normally two layers of glass with an interlayer of polyvinyl butyral (PVB) that holds the glass together when it is impacted. The most common use of laminated glass is in the windshields of automobiles. Tempered glass is a glass that has gone through a heating and cooling process in the manufacturing to strengthen it. When it breaks, it shatters into small fragments instead of large shards of glass. Tempered glass is used extensively in buildings due the lower cost than laminated glass. Safety glazing is required to be identified by a permanent identification mark. The mark can be acid etched, sand blasted, ceramic fired, laser etched, or embossed. Whatever type of mark is used, once applied, it cannot be removed without being destroyed. **[Ref. 2403.1]**

FIGURE 12-5 Safety glazing requirements at doors and adjacent to doors

Safety glazing is required in locations where the possibility of impacting the glass and getting seriously injured is high. The most common hazardous locations are doors and walls adjacent to doors. Glazing in doors including swinging doors, sliding and bifold doors, storm doors, and unframed swinging doors must be safety glass. Safety glazing is required in panels adjacent to a door where the nearest edge of the glazing is within a 24-inch arc of either vertical edge of the door in a closed position if the bottom edge of the glazing is less than 60 inches above the walking surface (see Figure 12-5). **[Ref. 2406.4]**

Another location where glass can be a hazard is when it is close to the walking surface and rather large in size. Four conditions need to be present to require a single panel of glass to be safety glazing. These requirements apply to both fixed and operable panels of glass:

1. The exposed area of an individual pane is greater than 9 square feet.
2. The exposed bottom edge of the pane is less than 18 inches above the floor.
3. The exposed top edge of the glass is greater than 36 inches above the floor.
4. There is one or more walking surfaces within 36 inches horizontal of the plane of the glazing.

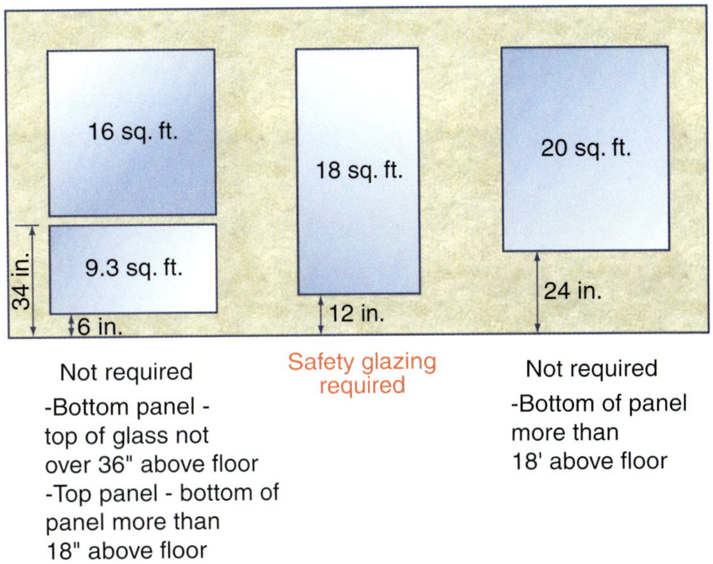

FIGURE 12-6 Safety glazing requirements

Remember that all four of the preceding conditions must be present to require the glass to be safety glazing. If just one of the conditions does not exist, the glass can be regular float glass (see Figure 12-6). **[Ref. 2406.4, 7]**

Bare feet and wet, slick surfaces do not go well with each other. Therefore, the IBC requires safety glazing in areas where people are walking or standing on a wet surface in either bare feet or shoes. Glass panels that are any part of an enclosure around a bathtub, shower, steam room, sauna, whirlpool, or hot tub must be safety glazing. This includes any doors used to access these areas (see Figure 12-7). If

FIGURE 12-7 Safety glazing required at windows next to tub and at glazing surrounding shower enclosure

the glass is more than 60 inches above the standing surface, it is not required to be safety glazing. Any glass in walls or fences enclosing hot tubs, spas, whirlpools, saunas, steam rooms, bathtubs, showers, and indoor or outdoor swimming pools must be safety glazing if the glass is less than 60 inches above the walking surface on the pool side of the glazing and it is within 60 inches horizontally of the waters edge. The 60-inch dimension is a common dimension applied to the location of safety glazing. When glazing is more than 60 inches away from a hazardous location, the likelihood of someone impacting the glass is reduced. **[Ref. 2406.4]**

FIGURE 12-8 Safety glazing required adjacent to stairways

The fact that stairs are an area where people fall has already been discussed. So, it only makes sense that any glass adjacent to a stairway be safety glazing. The code has three locations where glass next to a stairway is considered to be hazardous. (1) Glazing used for guards and railings including baluster panels and nonstructural in-fill panels must be safety glazing. (2) Glass that is within 36 inches horizontally and 60 inches of a walking surface of stairways, landings, and ramps must be safety glazing. (3) In addition, any glass located within 36 inches of the bottom tread of a stairway must be safety glazing if it is less than 60 inches above the floor (see Figures 12-8 and 12-9). **[Ref. 2406.4]**

FIGURE 12-9 Glazing used in guard required to be safety glazing

Two categories of safety glazing, Category I and Category II, are required by the IBC. Category II is a stronger glass than Category I. In the CPSC Standard 16 CFR 1201, Category I glass is impacted from a drop height of 18 inches and Category II is impacted from a drop height of 48 inches. Category II glass is required in all locations, except that Category I glass not exceeding 9 square feet is permitted to be used in the following locations:

- Glazing in storm or combination doors
- Glazing in doors
- Panels adjacent to doors when located as previously outlined

The ANSI Standard Z97.1 uses Category A and Category B to differentiate the strength of the safety glazing. Category A glass is required in locations previously identified except for those panels 9 square feet or less that are adjacent to doors. For locations not addressed in the IBC tables, such as glazing adjacent to pools or stairways, Class A or II glazing are acceptable. **[Ref. 2406.2]**

SWIMMING POOLS

Children love swimming pools. The pools are an attraction to them and they love to go play in them. Unfortunately, children are not always supervised and they sometimes fall into pools and drown. The U.S. Consumer Product Safety Commission reports that about 260 children under age five drown in swimming pools each year. Therefore, the building code has regulations to protect children from accessing pools without supervision and falling in. The code divides swimming pools into public and residential pools. A public swimming pool is considered to be any pool that is not associated with a Group R-3 single-family dwelling. Examples of public pools are municipal pools or those associated with apartments, hotels, or health clubs. **[Ref. 3109]**

A public swimming pool must be completely enclosed by a fence or screen enclosure that is a minimum of 4 feet in height. The openings in the fence have to be sized so a 4-inch sphere cannot be passed through the enclosure. Any gates that access the pool area must close automatically and self-latch (see Figure 12-10). **[Ref. 3109.3]**

Residential swimming pools have much more detailed requirements. The same intent of protecting children from accessing the pools is still there. However, different circumstances surround residential pools that are not necessarily present in public swimming pools. Residential pools must be surrounded with a barrier as well. However, the code allows the use of a power safety cover or a spa with a safety cover complying with ASTM Standard F1346. **[Ref. 3109.4]**

The barrier around a residential swimming pool is also required to be a minimum of 48 inches in height above grade measured on the side of the barrier that faces away from the pool. The maximum vertical clearance between the ground level and the bottom of the barrier is limited to 2 inches. Openings in the barrier must be sized so that a 4-inch-diameter sphere cannot be passed through the barrier. If the barrier is solid, it cannot contain any indentations or protrusions that may make it easy for a child to scale the wall and get into the pool area. Horizontal and diagonal members in barriers are also regulated so they cannot be used to climb them. If chain link is used as the barrier, the maximum mesh size is 2¼ inch squares. If the fence is provided with slats fastened at the top or bottom, a larger opening is permitted as long as the slats reduce the openings to 1¾ inches or less. The barriers must be located so that permanent structures, equipment, or similar objects cannot be used to climb up and over them.

Gates must also be constructed to restrict entry of children into the pool area. Doors and gates that serve pool areas must open outward away

> **Code Basics**
>
> **Swimming Pool**
>
> - Any structure intended for swimming, recreational bathing, or wading
> - Contains water over 24 inches deep
> - Includes in-ground, above-ground, and on-ground pools
> - Includes hot tubs and spas
> - Includes fixed-in-place wading pools

FIGURE 12-10 Public pool barrier

from the pool area. They must close by themselves and latch automatically. If the release mechanism is located below 54 inches from the bottom of the gate or door, it must be located on the pool side. It must also be located at least 3 inches from the top of the door or gate. The gate or door must be constructed so that it does not have any openings larger than ½ inch within 18 inches of the release mechanism. If the release mechanism is a self-locking device operated by a key, electronic opener, or integral combination lock, the device must be located between 34 and 48 inches above the floor per IBC Section 1008.1.9.3.

If the wall of the dwelling is used as part of the barrier, specific requirements must be met. Three options are available to comply with the code.

1. Doors that access the pool area directly from the dwelling must be equipped with an alarm that produces an audible warning when the door and/or its screen, if there is one, are opened. The deactivation switch must be located a minimum of 54 inches above the threshold of the door.
2. A power safety cover can be provided that complies with ASTM Standard F 1346.
3. Some means of protection, such as self-closing doors with self-latching devices, that provides the same degree of protection as the first two items can be approved.

Entrapment avoidance

Pool entrapment is when a portion of a person's body is sucked into the suction outlet in a pool water-circulation system. If this happens, the person can be seriously injured and possibly drown. New to the 2009 IBC is the requirement that swimming pools be provided with suction outlets that comply with ANSI/APSP-7 (APSP stands for the Association of Pool and Spa Professionals). Such an outlet cover protects against entrapment of a limb or body in the bottom of the pool. **[Ref. 3109.5]**

PART V

Health Safety

CHAPTER 13

Weather Protection

One of the biggest problems in buildings today is the intrusion of moisture. There has been an enormous amount of litigation on the installation of exterior wall coverings and roof systems. If these systems are not installed properly, the moisture can create a destructive reaction to the building and its contents. The moisture can also have a negative effect on the occupants in the building. This chapter discusses the requirements in the *International Building Code* (IBC) to mitigate moisture problems in a building.

EXTERIOR WALLS

Many different types of materials are used on the outside of buildings to protect them from moisture (see Figure 13-1). Chapter 14 of the IBC regulates the installation of exterior wall coverings, wall openings, windows and doors, architectural trim, and balconies and similar projections and openings. The IBC requires that the construction documents include a description of the exterior wall envelope in sufficient detail to determine compliance with this code. The construction documents must include details of the exterior wall envelope, including flashing; intersections with dissimilar materials; corners; end details; control joints; intersections at roof, eaves, or parapets; means of drainage; water-resistive membrane; and details around openings.

They must include the manufacturer's installation instructions that provide supporting documentation that the proposed penetration and opening details described in the construction documents maintain the weather resistance of the exterior wall envelope. The supporting documentation must describe the exterior wall system that was tested, where applicable, and the test procedure used.

Keeping the water out!

The best way to protect the building from moisture is to keep the water out of the building. The IBC requires that the building be provided with a weather-resistant exterior wall envelope including flashing. The wall must be designed to prevent accumulation of water within the wall assembly. This is accomplished by providing a water-resistive barrier behind the exterior veneer and a means of draining to the exterior any water that enters the assembly. It also requires that a vapor retarder be

FIGURE 13-1 Chapter 14 of the IBC regulates the installation of exterior wall coverings

provided on the interior side of framed walls to resist the passage of condensation into the wall assembly.

There are several ways to prevent water from entering a wall assembly. The primary concept is to provide a drainage plane away from the building. This is accomplished by installing the exterior wall covering materials in a proper manner. Flashing is the second important component of an exterior wall. This moves the water away from openings and dissimilar materials. There are hundreds of different materials used in the construction of an exterior wall assembly. These systems must be installed properly to meet the intent of keeping the water out of the exterior wall assembly.

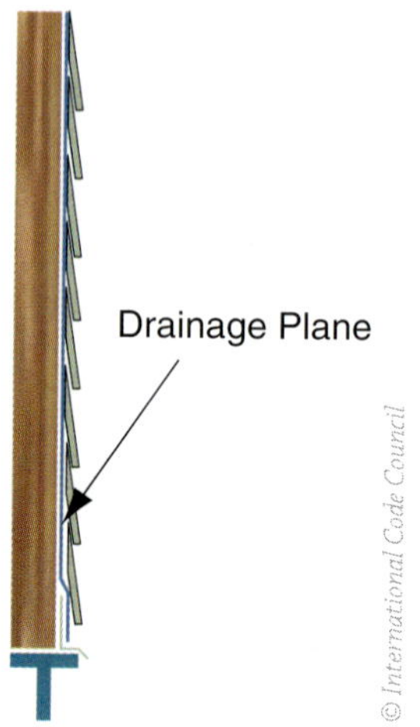

FIGURE 13-2 Drainage plane at exterior wall

> **Code Basics**
>
> **Flashing Required**
>
> - Perimeters or exterior doors and window assemblies
> - Penetrations and terminations of exterior wall assemblies
> - Exterior wall intersections with roofs, chimneys, porches, decks, balconies, and similar projections
> - Built-in gutters
> - Similar locations where moisture could enter the wall ●

Drainage plane

A drainage plane is the space between the exterior veneer and the water-resistive barrier behind the veneer. It must provide a direction for water to flow after it hits the side of a building and migrates through the wall covering. It is important that the drainage plane provide drainage down and away from the wall. The most common way to provide the drainage plane is to install weather-resistant materials in a shingle fashion over the material below. This material is typically installed behind the veneer (see Figure 13-2).

Flashing

Flashing is one of the most important components, other than the weather barrier, that must be installed in the exterior wall envelope. It prevents water from entering the wall at openings where building components penetrate the exterior wall assembly. The IBC requires flashing be installed in such a manner so as to prevent moisture from entering the wall or to direct it to the exterior. The code then specifies the location where flashing is required but does not specify how the flashing must be installed. There are too many different materials used for flashing systems. It would be impossible to include all of the different systems in the code. Therefore, the installation details of flashing are typically left up to the design professional and/or manufacturer of the flashing material. The instructions of the penetrating item, such as a door or window, must be followed as well (Figure 13-3).

Vapor retarders

A vapor retarder is a material, typically a plastic or foil sheet, that resists diffusion of moisture through the exterior envelope of a building. The building envelope is the exterior walls, ceilings, and floors of a building. Water vapor, created when water evaporates in the air, is produced in buildings in many ways. This vapor tends to move from the warm side of an assembly to the cool side. The codes have historically required vapor retarders on the "warm in winter" side of the thermal envelope. Recent research has shown that this may not be what is needed in some climates. Therefore, the IBC now requires the vapor retarder to be installed on the interior side of the thermal envelope of buildings in specific climate zones. A climate zone map of the United States was created by the U.S. Department of Energy and is located in the *International Energy Conservation*

FIGURE 13-3 Flashing provided around windows but not doorways

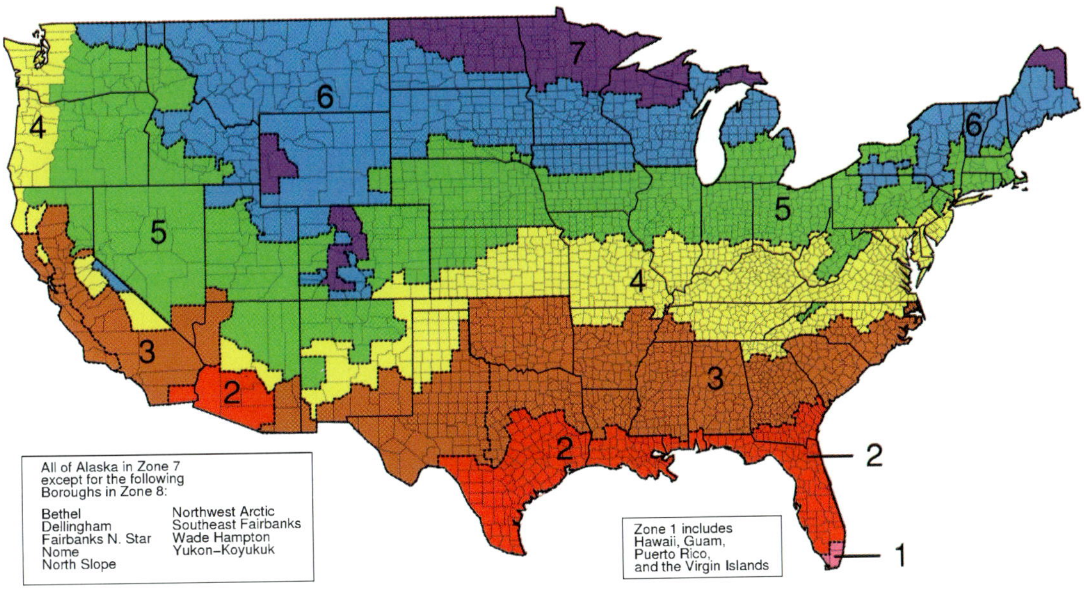

FIGURE 13-4 Climate zone map for the United States

Code (IECC) (see Figure 13-4). This map divides the country into different zones based on temperature, moisture, wind, and sun. It also accounts for the seasonal patterns of variations within these parameters. The IECC also has a list of counties in each state and in which zone they are located.

Three classifications of vapor retarders are referenced in the IBC: Class I, Class II, and Class III. Class I vapor retarders are typically sheet polyethylene and nonperforated aluminum foil. Class II retarders can either be kraft-faced fiberglass batt insulation or paint that has a perm rating greater than 0.1 and less than 1.0. Latex or enamel paint is considered to be a Class III vapor retarder. The IBC requires a Class I or Class II vapor retarder on the interior side of frame walls in Zones 5,

> **You Should Know**
>
> Permeability, rated in perms, is a measure of the rate of transfer of water vapor through a material (1 perm = 1 grain per hour, per square foot, per inch of mercury pressure). ●

6, 7, 8, and Marine 4. However, a vapor retarder is not required to be installed in basement walls, in a below-grade portion of any wall, or in construction in which moisture or its freezing will not damage the materials in the wall assembly. This is essentially the northern half of the United States. Class III vapor retarders can be used in specific zones based on the construction of the wall assembly. If vented cladding is used or insulated sheathing is installed, the use of latex or enamel paint could be used. Table 1406.3.1 of the IBC should be referenced for the specifics.

Materials

Water-resistive barrier

One of the last layers of protection from moisture entering the wall assembly is the water-resistant barrier. The IBC requires this barrier to be installed behind the exterior wall veneer. The barrier must be a minimum of one layer of No. 15 asphalt felt, complying with ASTM Standard D226 for Type 1 felt or other approved materials. The weather barrier must be installed in a shingle fashion and be properly lapped so the water drains away from the building. If it is not properly lapped, the water will be directed back into the wall and damage will occur (Figure 13-5).

A material called *building wrap* is typically used as an alternative approved material. Building wrap is a plastic-based sheet that prevents liquid water from passing through it. Another term used for this material is *stucco wrap.* It is important that the installation of the building wrap comply with installation instructions from the manufacturer. Manufacturers have specific requirements for lapping, fasteners, and terminations.

FIGURE 13-5 Building wrap water-resistant barrier

TABLE 13-1 Minimum thickness of siding (Based on IBC Table 1606.2)

Covering Type	Minimum Thickness (inches)
Aluminum siding	0.019
Fiber cement lap siding	0.25
Hardboard siding	0.25
Steel siding	0.0149
Vinyl siding	0.035
Wood siding (without sheathing)	0.5
Wood siding (with sheathing)	Permitted to be less than 0.5

Siding

Many different types of siding are available for exterior wall finishes. There are sidings made of wood, steel, aluminum, vinyl, hardboard, fiber cement, and other materials. The IBC regulates the minimum thickness of exterior materials. Table 13-1 outlines the minimum thickness of siding materials. All of the nonwood sidings are required to comply with specific standards. These standards test the durability and the capabilities of the material to resist wind-driven rain and snow, as well as provide a weather-resistant surface. These materials must also be installed in strict accordance with the manufacturer's installation standards. Wood siding must be nailed to the stud with 6d nails for ½ inch thick or less material and 8d nails for ⅝-inch material.

Masonry veneer

The two types of masonry veneer are anchored and adhered. Anchored masonry veneer is the traditional brick-type veneer in which the material is secured to a backing material (sheathing) by mechanical fasteners (see Figure 13-6). Adhered masonry veneer is veneer in which the material is secured and supported by the adhesion of a bonding material applied to the backing. It is important to remember that these veneers are not a structural element of the building. They are only an exterior finish material.

Anchored masonry veneer must be installed in accordance with The Masonry Society (TMS) *Building Code Requirements for Masonry Construction*. This standard provides the specific details and requirements for this type of material. In addition to the TMS standard, the IBC requires that flashing and weep holes be located in the first course of masonry above finished ground level, above the foundation wall or slab and other points of support (see Figure 13-7). Adhered masonry veneer is sometimes called thin-brick or faux-stone veneer. It is a relatively new type of material in the construction market. These veneers must be installed in accordance with the TMS standards as well as the manufacturer's installation standards.

Stucco

Stucco has become a very popular material for exterior wall finishes (see Figure 13-8). The IBC regulates the cement-based, three-coat stuccos. The new synthetic stucco systems are considered alternate materials and must be evaluated and approved by the building official prior to

FIGURE 13-6 Anchored veneer masonry

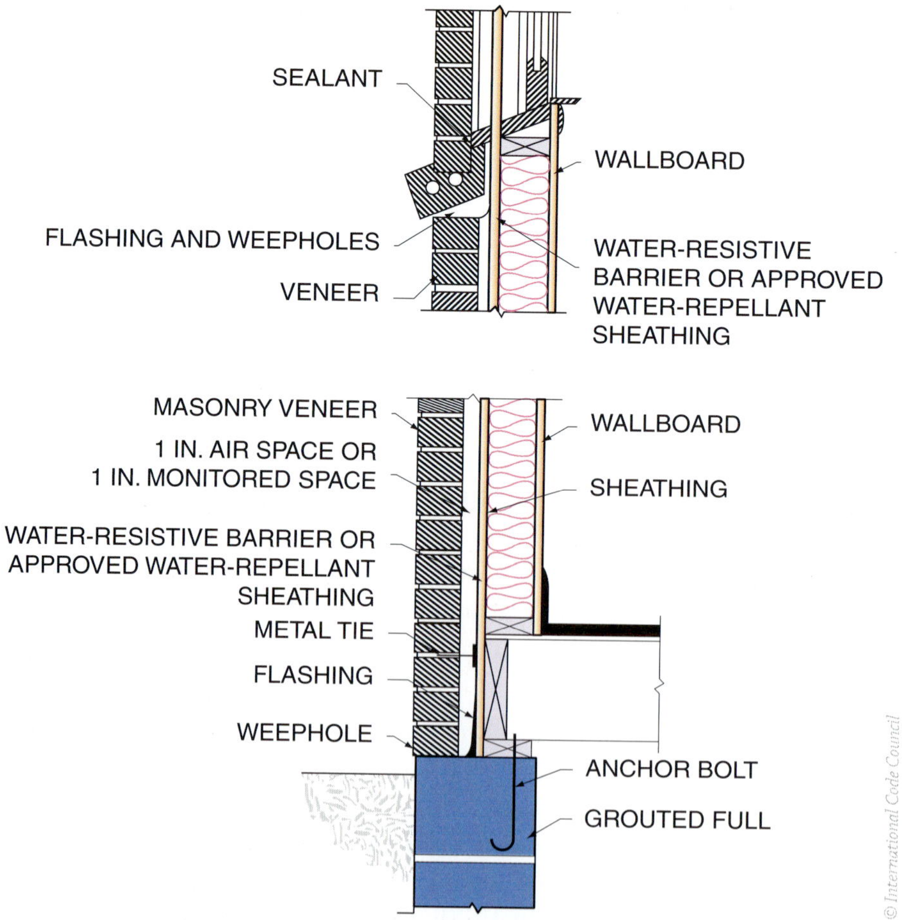

FIGURE 13-7 Masonry veneer detail

FIGURE 13-8 Stucco on exterior of office building

installation. The first difference between stucco and other exterior finishes is that if the material is installed over a wood based-sheathing, the water-resistive barrier must be doubled. Some types of building wraps are also required to be doubled. The installer should check the installation requirements of the system being used.

The installation of cement-based plaster (stucco) must comply with the provisions of IBC Section 2512. This section requires a weep screed at the base of the stucco to allow water to drain out from the weather barrier. It must be placed a minimum of 4 inches above ground and a minimum of 2 inches above paved areas. The weather barrier must overlap the attachment flange of the weep screed. The lath must also overlap the attachment flange.

ROOFING

The exterior wall assembly is not the only location at which water can get into a building. The roof system is also a prime location for water intrusion (see Figure 13-9). The IBC regulates the installation of roofing systems. The code requires that the roof coverings be installed in accordance with IBC Chapter 15 and the manufacturer's installation instructions. When there is a conflict between the code and the installation instructions, the code requirements must be met. The roofing system must be installed properly to keep water out of a building. The minimum slope of a roof is dependent on the type of roof material being used. The specific slope requirements are discussed later in this section.

The first thing that needs to be addressed is the slope of the roof. The code requires all roofs to be provided with drainage in accordance with the *International Plumbing Code* (IPC). The IPC provides a requirement

FIGURE 13-9 Metal panel roofing

on the sizing of the roof drain system and a requirement for secondary drains. Secondary drains or scuppers are required anytime the water cannot drain over the edge of a roof and water can build up if the primary drains are plugged. If secondary drains are not provided and the primary drains are plugged, the potential of a roof collapsing increases.

Flashing

Just like the exterior wall assemblies, flashing is an important component of a roofing system. It provides protection from leaks at changes of direction of the roof surface. Flashing must be installed so it prevents moisture from entering walls and roofs through joints in copings, through moisture-permeable materials, and at intersections with parapet walls and other penetrations through the roof. Flashing must be installed at wall and roof intersections, at gutters, wherever there is a change in roof slope or direction, and around roof openings. Parapets must be properly covered (coped) with noncombustible, weatherproof materials that are the same width as the thickness of the parapet wall (Figure 13-10).

Many different materials are used for flashing on a roof. When metal is used as a flashing, it must be corrosion resistant and a minimum of No.-26-gauge galvanized sheet metal. Other flashing materials must be installed in accordance with their manufacturer's installation instructions. Whatever material is used for the flashing must be compatible with the rest of the materials used in the roofing system.

Low-slope roofs

The term *flat roofs* is a little misleading in that they are really not flat. They have a very low slope and may look flat to the naked eye. Low-slope roofs typically have a minimum slope of ¼-inch vertical to 12-inch horizontal (¼:12 or 2 percent) depending on the roofing material. Roofs are considered low-slope roofs up to 3:12 pitch.

FIGURE 13-10 Low-slope roof flashing

Low-slope roofs traditionally have been constructed with an asphalt built-up roof covering. This system is a series of multiple layers of asphalt-impregnated sheets of felt paper and hot asphalt. Most built-up roof systems are comprised of three layers of felt and asphalt. The top of the roof system is then covered with a gravel surface to protect the asphalt from ultraviolet (UV) rays of the sun. The UV rays degrade the oils in the asphalt and reduce the effectiveness of the roof system. Built-up roofs are required to be a minimum of ¼:12-inch slope. When coal tar pitch is used in a built-up roof, the roof slope is permitted to be reduced to ⅛:12 inch. The materials in these systems must comply with several ASTM standards.

Different materials have been introduced into the market for low-slope roofing. All of these systems must be installed in accordance with the manufacturer's installation instructions. Most of these materials are single-ply systems, which means that there is only one layer of membrane material applied as opposed to a built-up roof or multiple-ply system. Modified Bitumen roof coverings are one or more layers of polymer-modified asphalt sheets. These systems can be either fully adhered to the roof sheathing or mechanically fastened. In some cases, these systems are held in place by a ballast of large stones or paver systems. The ballast resists uplift of the membrane from wind passing over the top of the roof and creating a negative pressure on the system (Figure 13-11).

Two other single-ply roofing membranes are Thermoset and Thermoplastic single-ply roofing systems. Thermoset single-ply roofing is a synthetic rubber sheet that is adhered together with contact adhesive or tape. These systems can be adhered, mechanically fastened, or held down with a ballast. They are sometimes referred to as *EPDM roofing*

FIGURE 13-11 Ballasted single-ply roof system

FIGURE 13-12 Thermoset single-ply adhered roof system

(see Figure 13-12). Thermoplastic single-ply roofing systems are plastic sheets (PVC, TPO, CSPE) that are welded together with hot air. This creates a shingle sheet that provides the weather protection for the building. Both the Thermoset and the Thermoplastic must be installed at a minimum ¼:12-inch slope.

Steep roofs

Roofs with a greater than 3:12 slope or pitch are considered to be steep roofs. Common materials used in steep roofing include asphalt shingles, wood shakes or shingles, and clay or concrete tiles. The IBC provides prescriptive installation requirements for these three systems as well as metal roof panels and shingles. In all of these roof systems, there are five important components: decking/sheathing, underlayment, membrane, fastening, and flashing. Underlayment is one or more layers of asphalt-impregnated felt, sheathing paper, or nonbituminous saturated felt. It is applied over the decking material prior to the installation of a roof membrane. The code outlines the installation requirements for each of these components depending on the type of system being installed.

Asphalt shingles

Asphalt shingles must be installed on solidly sheathed decking. This is typically plywood or some other type of structural sheathing. The shingles must be installed on slopes of 2:12 inches or greater. However, if they are installed on roofs with a slope between 2:12 and 4:12, two layers of underlayment are required. The double layer of underlayment must be overlapped a minimum of 19 inches and fastened to roof sufficiently to hold it in place before the shingles are installed. The single-layer underlayment is required to be lapped a minimum of 2 inches. In areas of the country where there has been a history of ice-damming along the eaves, additional layers of an ice barrier must be installed along the eave of the roof extending up the roof to a point 24 inches inside the exterior wall line of the building.

Flashing is required to be installed along the roof and wall intersections, in valleys, and along the drip edge of the roof. Base flashing is the L-shaped flashing that is attached to the roof itself. It extends under the roofing membrane with the other leg then extending up the adjacent wall. A cap flashing is then installed over the base flashing (see Figure 3-13). If siding is used, the base flashing is extended up behind the siding and a cap flashing is not installed. Valley flashing is installed where two different roof slopes come together and form a "valley" down the roof. Asphalt shingles have two different types of valleys, closed and open. The code provides specific requirements on how these valleys must be flashed. Drip-edge flashing is provided at the edges of the eaves and gables of asphalt shingle roofs.

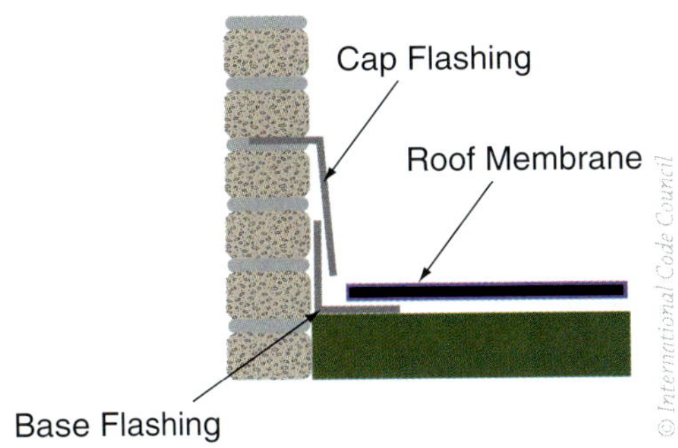

FIGURE 13-13 Cross section of base and cap flashing

The shingles must be fastened to the decking with galvanized, stainless steel, aluminum, or copper roofing nails. They must penetrate through the shingles and a minimum of ¾ inch into the roof sheathing, or through it if it less than ¾-inch thick. A minimum of four nails or the minimum number of fasteners required by the manufacturer must be provided per strip of shingles. In areas with high winds, additional nailing may be required. The asphalt shingles must also comply with specific ASTM standards in high-wind areas.

FIGURE 13-14 Clay tile roof system

Clay and concrete tiles

Clay and concrete tiles can be installed over solid sheathing or spaced structural sheathing boards (see Figure 13-14). The minimum slope allowed for tiles is 2½:12. A double underlayment must be provided when the slope is between 2½:12 inches and 4:12. Tile roofs that are steeper than 4:12 only require one layer of underlayment. The underlayment must be a Type II asphalt felt or Class M mineral-surfaced roll roofing. The ice barrier noted earlier is also required for tile roofs when ice damming is a possibility.

The tile must comply with two ASTM standards. Clay tile must comply with ASTM Standard C1167, and concrete tile must comply with ASTM Standard C1492. Both must be fastened to the roof decking in accordance with IBC Table 1507.3.7, which outlines the fasteners required for each type of roofing tile based on the wind speed, height of the roof, and slope of the roof. The fasteners must be corrosion resistant and penetrate the decking a minimum of ¾ inch or through the deck if it is less than ¾-inch thick.

The roof tile must be installed in accordance with the manufacturer's installation requirements based on the climatic conditions, roof slope, underlayment system, and type of tile. Flashing must be installed at the juncture of the roof and wall surfaces, valleys, and penetrations through the roofing material.

The exterior envelope is the first line of protection for the building from the elements and moisture. Water that is allowed to enter wall and roof cavities can create serious damage to the building and its contents. It is important that exterior wall and roof systems be installed in accordance with the code requirements and the installation instructions from the roof system manufacturer. The proper installation of these systems will protect the building for many years to come.

CHAPTER 14

Interior Environment

© International Code Council

For people to be healthy, they must have certain basic things. We need adequate light to work or live in a building. We need fresh air that is free from contaminants. When it is cold, we need to be provided with heat to keep from getting sick. We also need freshwater and sanitary waste facilities. This chapter discusses the *International Building Code* requirements for providing a healthy environment in which to live and work.

LIGHT AND VENTILATION

Providing the required light and ventilation is discussed in this section. Lighting requirements include the use of both natural means and artificial means of illuminating a room or space. The IBC requires natural ventilation to be provided. However, when such ventilation cannot be provided, mechanical ventilation designed and installed in accordance with the *International Mechanical Code* (IMC) is permitted to be provided. Ventilation of attics and crawl spaces is also discussed.

Natural and artificial lighting

People need light to live and work, but they also need light to be healthy. Therefore, the IBC requires a minimum level of lighting in a room or space. The code provides an option for the use of either natural or artificial lighting. Natural lighting is provided by openings, typically windows, in exterior walls that allow the sun to shine into the building. Artificial lighting is the electrically powered lighting provided in a building.

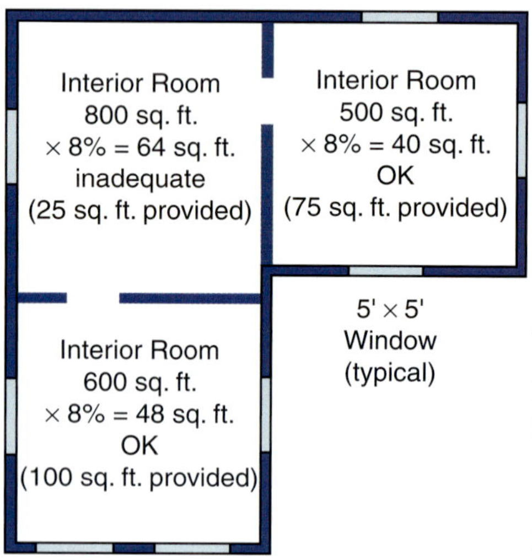

FIGURE 14-1 Natural light calculation

To provide the natural light, the code requires a minimum net exterior glazed area of openings. The exterior openings must be at least 8 percent of the floor area of the room or space that they serve (see Figure 14-1). These required openings must open onto a public way, yard, or court. Public ways are streets, alleys, or other parcels of land that lead to a public right-of-way. Courts are open spaces that are surrounded by exterior walls of the building on at least three sides. Yards are open spaces that are unobstructed to the sky and located on the same lot as the building. The minimum dimensions of courts and yards are dependent on the height of the building. As the building gets taller, the yard or court must be increased. **[Ref. 1205.2]**

When there is inadequate natural light in a room or space, as noted in the upper-right-hand room in Figure 14-1, the IBC permits adjoining spaces to share the natural light. In order to use this option, there must be an opening in the common wall that is at least 50 percent of the area of the wall. The opening in the common wall must also be at least 10 percent of the floor area of the interior room or 20 square feet, whichever is greater (see Figure 14-2). **[Ref. 1205.2.1]**

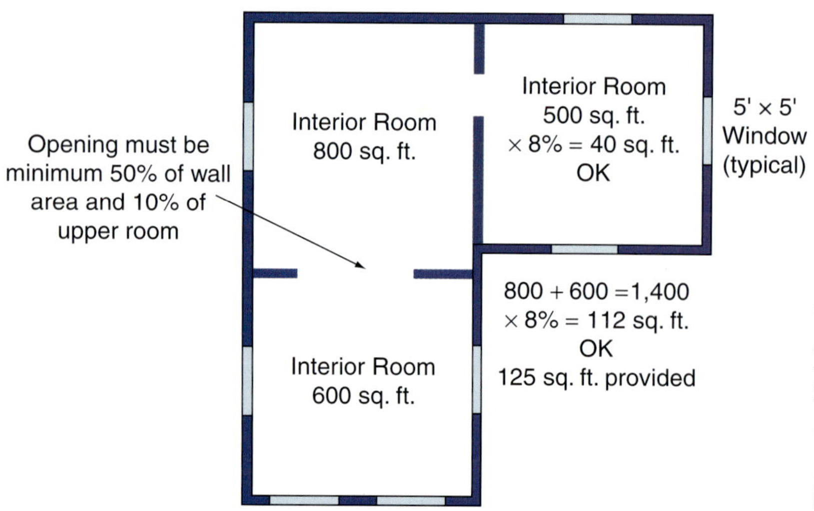

FIGURE 14-2 Shared natural light calculation

If natural light cannot be provided for a room or space, the code permits the installation of artificial light. The lighting must provide an average illumination of ten foot-candles over the area of the room at a height of 30 inches above the floor. The level of lighting can be checked using a light meter held at the required height.

Even though natural light is rarely used to meet the code requirements due to the presence of artificial lighting, the use of natural light is an option to the designer. This requirement was originally included in many of the early codes and has survived over the years as a requirement in the code. By providing natural light, the use of artificial lighting can be reduced and energy savings incurred. **[Ref. 1205.3]**

Natural ventilation

Natural ventilation can be provided in many ways. It can be provided through windows, doors, louvers, or any other openings to the outdoors. The openings must be operable by the building occupants and be readily accessible. There are no requirements that they be held open, just that they are openable.

The minimum openable area of ventilation openings must be 4 percent of the floor area being ventilated. Therefore, if the window openings shown in Figures 14-1 and 14-2 were windows that opened at least 50 percent, they would comply with this requirement. The code also allows rooms and spaces without openings to be ventilated through an adjoining room similar to the shared lighting discussed earlier. However, the size of the wall opening between the two rooms must be a minimum of 8 percent of the floor area of the interior room or spaces, but not less than 25 square feet. The area of the ventilation openings is based on the total area of both rooms. As with the natural light provisions, the openings for natural ventilation must open into yards, courts, or public ways (see Figure 14-3). **[Ref. 1203.4]**

FIGURE 14-3 Natural ventilation openings used on second floor of mixed-used building

An option to natural ventilation is the use of mechanical ventilation. The design and installation of a mechanical system must be in accordance with the IMC. The IMC requires a minimum amount of outside air to be introduced into the building as a part of the building's heating and/or air-conditioning system. The amount of outside air is based on the use or occupancy of the building. For example, more outside air is required in assembly occupancies than in office spaces, due to the number of people that may be in the building.

Exhaust systems

Excess moisture in buildings can create health problems if not exhausted. One of the largest contributors of moisture is bathrooms. Therefore, the IBC requires that bathrooms that contain bathtubs, spas, and similar bathing fixtures be mechanically ventilated in accordance with the IMC. This is typically accomplished with an exhaust fan, which must be extended to the exterior of the building. **[Ref. 1203.4.2.1]**

Contaminants must also be removed from buildings to maintain a healthy environment. Therefore, the IBC requires that contaminant sources in naturally ventilated spaces be removed in accordance with the IMC and *International Fire Code* (IFC). Any type of flammable and combustible hazards must also be exhausted in accordance with the IMC and IFC. **[Ref. 1203.4.1 and 1203.5]**

TEMPERATURE

To maintain a minimum level of heat in a building, the IBC requires that any interior space intended for human occupancy be provided with a space-heating system. The system is permitted to be either an active or passive system to address new technologies in heating. The system must be capable of maintaining a minimum indoor temperature of 68° F at a point 3 feet above the floor on the design heating day. The *design heating day* is the outdoor temperature used to design the heating system. These temperatures can be found in Appendix D of the *International Plumbing Code* (IPC) under the Winter Design Temperatures column (see Table 14-1). The code does not require that the heating system operate constantly. It only mandates that a system be provided that is capable of maintaining the minimum required temperature. It also does not require air conditioning within a building. **[Ref. 1204]**

TABLE 14-1 Examples of design heating days (Based on IPC Table D101)

Location	Winter Design Temperature (degrees F)
Minneapolis, MN	–12
Chicago, IL (O'Hare)	–4
Buffalo, NY	6
Kansas City, MO	6
Dallas, TX	22
San Diego, CA	44

MOISTURE CONTROL

As noted in Chapter 13, moisture within a building can become a problem. The condensation created within a building and emitted from the ground underneath a building must be allowed to escape. The IBC provides a path for the moisture to leave these spaces by ventilating the attic and under-floor spaces.

Attic spaces

Attic spaces hold in moisture that is created within a building. Therefore, the code requires that enclosed attics and rafter spaces where ceilings are applied directly to the underside of the framing members be ventilated. Cross ventilation must be provided for each separate space by openings to the exterior. The openings have to be protected against the entrance of rain and snow. Any insulation or blocking must be installed so that it does not obstruct the flow of air. A minimum of 1 inch of air space is required between the insulation and bottom of the roof sheathing.

The net free area of attic vents must be at least 1/150 of the area of the attic space being ventilated. There are two exceptions that allow vents sized at 1/300 of the attic area. The first exception can be used when 50–80 percent of the ventilation openings are located in the upper portion of the attic or rafter space and at least 3 feet above eave or cornice vents. The balance of the ventilation openings are provided by eave or cornice vents. This creates a natural air flow from the eave or cornice vents upward and out of the upper ventilation openings (see Figure 14-4). The other exception permits the reduced ventilation when a Class I or II vapor retarder is installed on the warm-in-winter side of the ceiling. A third option allows the building official to make the decision that the ventilation is not required if atmospheric or climatic conditions are such that attic ventilation is unnecessary. **[Ref. 1203.2]**

Under-floor ventilation

A crawl space under a building is a good space for mold and mildew to grow if there is not adequate ventilation for the space. The code requires any space between the bottom of the floor joists and the earth below, other than basements or cellars, to be provided with ventilation. The moisture in the ground is released into the under-floor space and trapped. If there is any type of cellulose material in the space, there is a likelihood that mold will begin growing. Ventilation openings are required to be placed through the foundation or exterior walls. They must be located so that they will provide cross ventilation of the space.

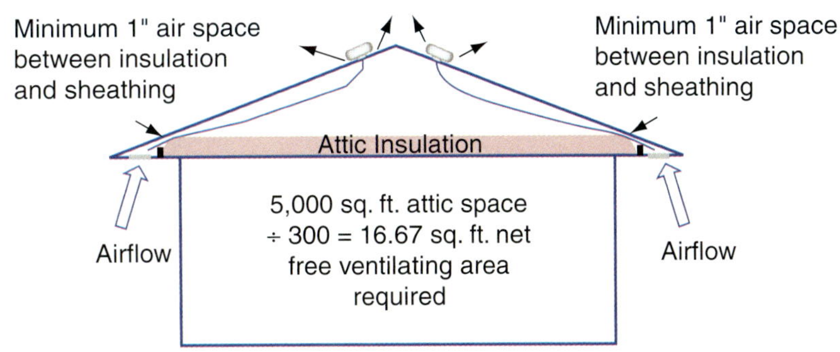

FIGURE 14-4 Attic ventilation requirements

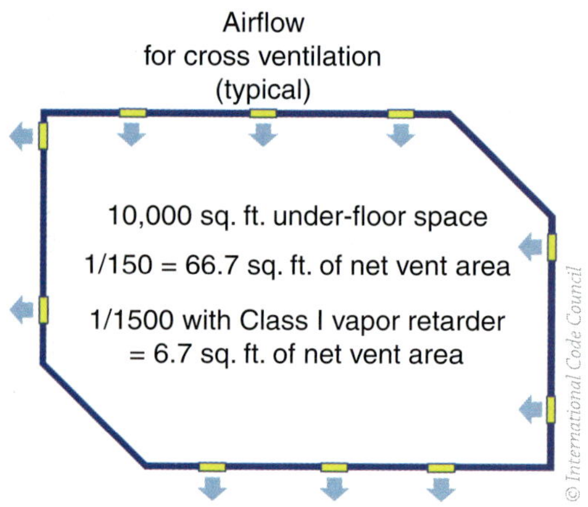

FIGURE 14-5 Under-floor ventilation requirements

The openings must be sized at a ratio of 1 square foot of opening for every 150 square feet of under-floor space. If a Class I vapor retarder material is provided over the ground and the openings are not provided with louvers, the ratio may be reduced to 1 square foot of vent opening for every 1,500 square feet of under-floor area (see Figure 14-5). **[Ref. 1203.3]**

PLUMBING FACILITIES

The United States has one of the safest plumbing systems in the world. Much of this is attributed to the code requirements that have been developed over the years. To provide a safe and sanitary system in a building, the IBC regulates the number, location, and privacy requirements for plumbing facilities. The installation of the plumbing systems must comply with the *International Plumbing Code* (IPC).

Minimum plumbing facilities

The minimum number of plumbing facilities is based on the number of people in a building. Table 14-2 provides a few examples of the number of facilities required in IBC Table 2902.1. The number of occupants in the building is determined by the IBC. The occupant load calculated for the building is typically used. The total occupant load is divided in half to determine the number for each sex. If approved statistical data are provided showing that the distribution of sexes is not 50 percent of each sex, the ratio can be revised. The number of plumbing facilities is calculated by using the fixture ratios shown in the table. Urinals may be substituted for water closets at a maximum of 67 percent for assembly and educational occupancies and no more than 50 percent for all other occupancies. This portion of the code originates in the IPC and is duplicated in the IBC. **[Ref. 2902.1]**

When plumbing fixtures are required, separate facilities must be provided for each sex in most buildings. Dwelling units in apartment buildings and sleeping units in hotels do not have to have separate facilities. Separate facilities are also not required in tenant spaces that have a total occupant load of fifteen or less, including both employees and customers, and in mercantile occupancies that have an occupant load of 100 or less. **[Ref. 2902.2]**

Code Basics

Given: 10,000 square foot office

Determine: Total required plumbing facilities

Solution:

Occupant load = 10,000 ÷ 100 = 100

100 ÷ 2 = 50 male, 50 female

Facilities Required:

Male toilet room

- Two water closets or one urinal and one water closet
- Two lavatories

Female toilet room

- Two water closets
- Two lavatories

One drinking fountain

One service sink ●

Location of toilet facilities

Toilet facilities located in occupancies that are generally open to the public must provide the facilities to all users including customers, patrons, and visitors. Employees must also have access to plumbing fixtures at all times the building is occupied. The employee facilities must either be separate or combined employee and public toilet facilities. Access to the public toilet facilities is not permitted to pass through kitchens, storage rooms, or closets. The route must also be accessible, as discussed previously in Chapter 11.

TABLE 14-2 Minimum number of required plumbing fixtures (Based on IBC Table 2902.1)

Classification	Occupancy	Description	Water Closets		Lavatories		Bathtubs/ Showers	Drinking Fountains*	Other
			Male	Female	Male	Female			
Assembly	A-2	Restaurant	1 per 75	1 per 75	1 per 200		–	1 per 500	1 service sink
	A-3	Auditoriums without permanent seating, art galleries, exhibition halls, museums, lecture halls, libraries, arcades, and gymnasiums	1 per 125	1 per 65	1 per 200		–	1 per 500	1 service sink
Business	B	Office	1 per 25 for the first 50 and 1 per 50 for the remainder exceeding 50		1 per 40 for the first 80 and 1 per 80 for the remainder exceeding 80		–	1 per 100	1 service sink
Educational	E	Educational facilities	1 per 50		1 per 50		–	1 per 100	1 service sink
Factory and Industrial	F-1 and F-2	Structures in which occupants are engaged in work fabricating, assembly or processing of products or materials	1 per 100		1 per 100		See Section 411 of the International Plumbing Code	1 per 400	1 service sink
Mercantile	M	Retail Store	1 per 500		1 per 750		–	1 per 1,000	1 service sink
Residential	R-2	Apartment House	1 per dwelling unit		1 per dwelling unit		1 per dwelling unit	–	1 kitchen sink per dwelling unit; 1 automatic clothes washer connection per 20 dwelling units
Storage	S-1 and S-2	Structures for the storage of goods, warehouses, storehouses and freight depots, low and moderate hazard	1 per 100		1 per 100		See Section 411 of the International Plumbing Code	1 per 1,000	1 service sink

In buildings other than covered mall buildings, the required toilet facilities for the public and/or employees must be located on not more than one story above or below the space required to have the facilities. The maximum distance a person must travel to get to the facilities is not allowed to exceed 500 feet. Longer distances are permitted in factory and industrial occupancies when approved by the building official. The required plumbing facilities must be provided free of charge. If pay facilities are provided, they must be in excess of the required facilities. **[Ref. 2902.3]**

Toilet/bathing room finishes

To maintain a sanitary and cleanable surface in toilet and bathing rooms, the IBC regulates the type of finishes used on the walls and floors. The floor finish in toilet, bathing, or shower rooms must have a smooth, hard, nonabsorbent surface. The vertical base installed at the intersection of such floors and walls must also be smooth, hard, and nonabsorbent and must extend up the wall at least 4 inches. The walls and partitions within 2 feet of service sinks, urinals and water closets must be covered with a smooth, hard, nonabsorbent surface. This surface must extend to a height of 4 feet above the finished floor. The most common way to meet these requirements is to install ceramic tiles on the floors and walls of the toilet rooms (see Figure 14-6). The materials used in the wall assembly must be of a type that is not adversely affected by moisture. Moisture-resistant gypsum board is a common material used in these locations. Shower compartments must also be finished with a smooth, nonabsorbent surface to a height of 70 inches above the drain inlet. **[Ref. 1210]**

FIGURE 14-6 Ceramic tile next to and behind water closet and extending up 4 feet above floor. Also note tile flooring

FIGURE 14-7 Privacy panels installed between urinals

In addition to the wall finish materials, privacy partitions must be provided in toilet facilities. Each water closet must be enclosed in a separate compartment with walls or partitions and a door to ensure privacy. If the toilet facility is a single-occupant toilet room with a lockable door, a separate compartment is not required. In addition, urinals must be provided with privacy panels. The panels must begin at a height not more than 12 inches from and extend not less than 60 inches above the finished floor surface. They must extend from the wall surface at each side of the urinal a minimum of 18 inches or to a point not less than 6 inches beyond the outermost front lip of the urinal (see Figure 14-7). **[Ref. 2903]**

PART VI Structural Safety

© International Code Council

CHAPTER

15 Structural Design

© International Code Council

Many structural forces are placed on a building over the intended life of the structure. Natural or environmental forces, as well as man-made loads, are placed on the building. There are basic design parameters outlined in the code for the design of a structure. These design parameters provide a minimum standard to ensure that the building withstands the forces applied to it. This chapter discusses how the *International Building Code* regulates the structural design of the building to ensure that it remains stable and does not collapse during construction and as long as the building is used. This chapter discusses the basic premise of the way buildings are designed to resist the loads applied to them.

Buildings are generally designed in accordance with the American Society of Civil Engineers (ASCE) Standard 7, Minimum Design Loads for Buildings and Other Structures. This document outlines the engineering calculations and methods of designing a building. Along with the building code, it evaluates the building based on its importance to the community. For example, a hospital that has emergency treatment facilities is essential in a disaster situation. Therefore, the structure for this building must be designed with higher safety factors than a typical office building. Four levels of risk categories are used in the structural design of a building: Category I, Category II, Category III, and Category IV. Category IV occupancy, which includes a hospital with emergency treatment facilities, has the highest importance factor and safety factors. Category I occupancy includes structures like agricultural buildings and minor storage facilities. Most buildings are classified as Category II risk. Category III occupancies include assembly occupancies having an occupant load greater than 300 and small hospitals without emergency treatment facilities. **[Ref. Table 1604.5]**

BASIC LOADS

There are many types of loads imposed on a building. The two main imposed loads used in the design of buildings are gravity loads and lateral loads. *Gravity loads* are those loads applied vertically to the structure. *Lateral loads* are applied in a horizontal manner to the building. Vertical loads include such loads as live loads, dead loads, and snow loads. Lateral loads are loads such as wind loads, earthquake loads, and flood loads. These loads are applied in two different ways to the building, uniform loading and concentrated loading. *Uniform loading* means that the load is applied equally across the structural member. An example of a uniform load is the floor framing that bears across the entire span of a beam. A *concentrated load* is applied in a single location of the member like a column bearing in the middle of a beam. A concentrated load is assumed to be applied uniformly over an area of 2½ feet by 2½ feet. In some cases, the concentrated loads and uniform loads are applied to the same member (see Figure 15-1).

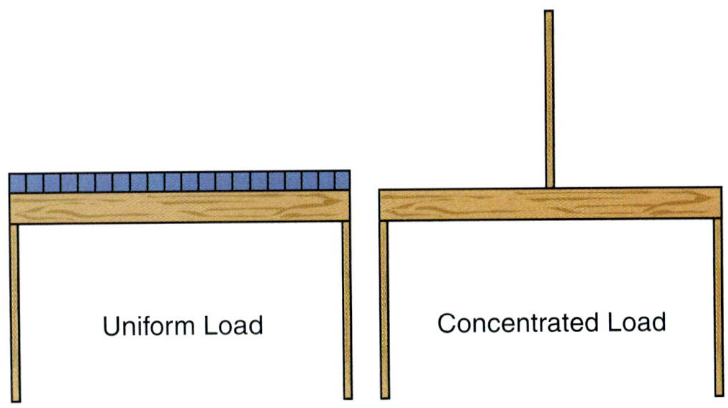

FIGURE 15-1 Uniform and concentrated loads

Live loads

Live loads are the loads created by the use and occupancy of the building. This is the weight of the furniture, equipment, people, and any other loads that may come and go on a structural member. Roofs also have a live load, which includes the weight of the workers, equipment, and materials used for the maintenance of the roof and/or equipment located on the roof. It also includes movable objects that may be placed on the roof like planters or people. The code provides a table of common live

TABLE 15-1 Common live loads used in buildings (Based on IBC Table 1607.1)

Occupancy or Use	Uniform (psf)	Concentrated (lbs.)
Dining rooms and restaurants	100	–
Office spaces	50	2,000
Residential 1- and 2-family dwellings		
Habitable attics and sleeping areas	30	
Other areas	40	–
Retail		
First floor	100	1,000
Upper floors	75	1,000
Stairs and exits		300
1- and 2-family dwellings	40	
All others	100	
Storage warehouses		
Heavy	250	
Light	125	
(must be designed for heavier loads if required for anticipated loads)		
Roofs		
Roof surfaces subject to maintenance workers		300
Ordinary flat, pitched, and curved roofs	20	
Occupiable roofs	100	

loads used in the design of a building (see Table 15-1). Live loads are designated in pounds per square foot (psf). Therefore, it is assumed that the weight of the people, furniture, and materials will be distributed evenly across the horizontal assembly. Concentrated loads are designated in pounds (lbs.). **[Ref. 1607]**

Dead loads

Dead loads are the actual weight of the materials used in construction of the building. They are considered permanent loads because they will always exist in a building. When designing the structural portion of a building, the design professional must use the actual weight of materials. If the information is not available, the building official must approve the design loads used. Many different resources are available outlining the weights of different materials in buildings. Typical values for weights of materials are shown in Table 15-2.

Deflection

Deflection is the amount that a structural member sags under the specified load without failing. It is a design factor that determines how stiff the member is. This is one of the many requirements that must be met within the structural design of the building. The code regulates the maximum deflection permitted for different structural members. Deflection limits are specified for roof members, floor members, and interior and exterior walls. The maximum deflection permitted for these different members is shown in Table 15-3. **[Ref. 1604.3]**

TABLE 15-2 Design dead loads (Based on Table C3-1 of ASCE 7)

Component	Load (psf)
Double wood floor supported on wood joists 16-inch on center	7
Hardwood flooring, 7/7-inch	4
Linoleum tile	1
Wood stud walls with ½ inch gypsum board each side	8
5/8 inch gypsum board ceiling	9
Asphalt shingles	2

TABLE 15-3 Deflection limits (Based on IBC Table 1604.3)

Construction	Maximum Deflection Under: Live Load	Wind or Seismic Load	Total Load
Roof members:			
Supporting plaster or stucco ceiling	l/360	l/360	l/240
Supporting nonplaster ceiling	l/240	l/240	l/180
Not supporting ceiling	l/180	l/180	l/120
Floor members	l/360	–	l/240
Exterior walls and interior partitions:			
With stucco or plaster finishes	–	–	1/360
With brittle finishes	–	l/240	–
With flexible finishes	–	l/120	–

The deflection is measured in inches. The length of the structural member such as a beam is also measured in inches and then divided by the maximum deflection limit to determine its stiffness. Figure 15-2 shows the deflection of a member and how the deflection is calculated. If the deflection of a member exceeds these limits, the member is not stiff enough. Most materials will return to their original state after they have deflected no more than the amount permitted by the code. Materials that deflect more than the limitations may not return to their original position and are then determined to have failed structurally.

The deflection limits shown in Table 15-3 are in some cases dependent on the type of materials applied to the structural member. The more brittle the material, the stiffer the member must be. If a brittle finish such as plaster is used, the member must be stiffer to prevent the cracking of the finish. If a nonbrittle finish is used, the member can deflect more and not affect it.

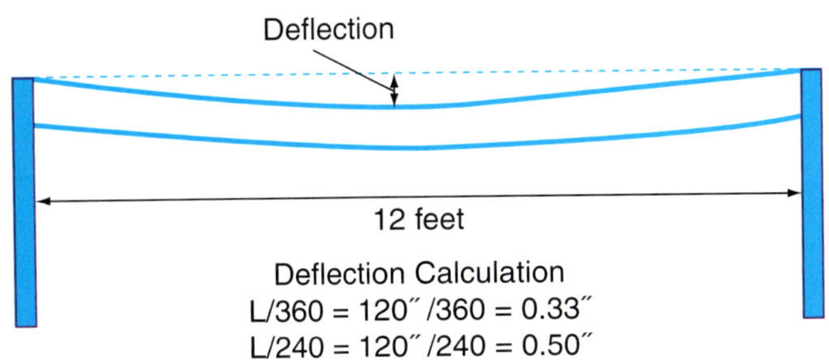

FIGURE 15-2 Deflection limitations

FIGURE 15-3 36 inches of snow creates heavy snow load in Colorado

Environmental loads

Several different types of environmental loads can be applied to a building. The most common ones are snow, earthquake, wind, and flood loads. Snow loads are vertical loads, and the others are considered lateral loads (see Figure 15-3). All of these loads must be taken into account in the structural design of the building depending on where the building is located.

Snow loads

The snow load is the weight of the snow that accumulates on a roof surface. The design snow loads in the IBC are based on the ground snow load specified on the map of the continental United States in Figure 15-4. The snow loads in Alaska are stated in IBC Table 1608.2. Hawaii's snow load is zero, except in mountainous regions as approved by the building official. These snow load values are based on 40 years of ground snow load data. They also include consideration of other factors involved with the buildup of snow on a roof. The thermal resistance of the roof structure can have an effect on the amount of snow that melts off the roof. There are also conditions in which the snow can create drifts or ponding that can create an uneven load on the structural member.

The snow loads are sometimes determined by the elevation. The numbers in parentheses refer to the upper elevation at which the snow loads occur. The areas on the map designated "CS," such as the western half of Colorado, are designated as case study areas. The snow loads in these areas must be determined by the local jurisdiction. Ground snow loads in these areas can range from one hundred to four hundred pounds per square foot. The code does allow the ground snow loads to be reduced due to the pitch of the roof and the fact that the wind will blow some of it off. However, the uneven loads of drifting must be taken into account in the structural design. [Ref. 1608]

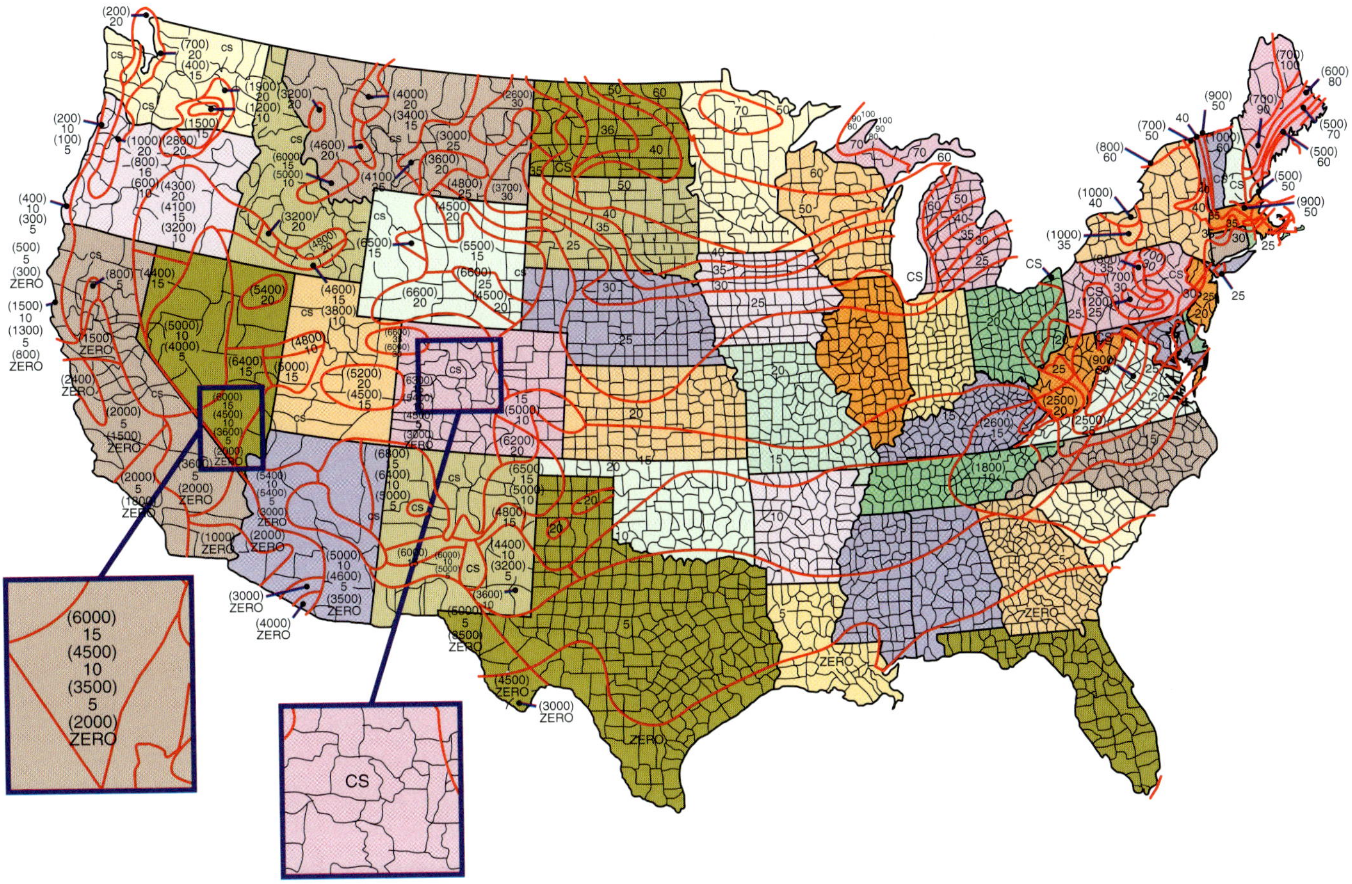

FIGURE 15-4 Ground snow loads for the United States (based on IBC Figure 1608.2)

Wind loads

The wind load is the pressure applied against the structure from winds. Wind loads do not act the same as the gravity loads already discussed. They generally act laterally against the building. However, they can create both positive and negative pressures on a building. When the wind blows directly against an exterior wall, it acts positively against the wall trying to push against it. This is called a positive pressure on the building. However, when the wind hits the corner of a building or the top of a roof, it creates suction on the opposite side that is called a negative force on the building (see Figure 15-5).

Wind loads for the different risk category buildings are determined based on the speed of the wind. Figure 15-6 shows the various wind speeds for risk Category II buildings. There are separate wind speed maps for risk Categories I, III, and IV. The wind speeds are equivalent to a three-second

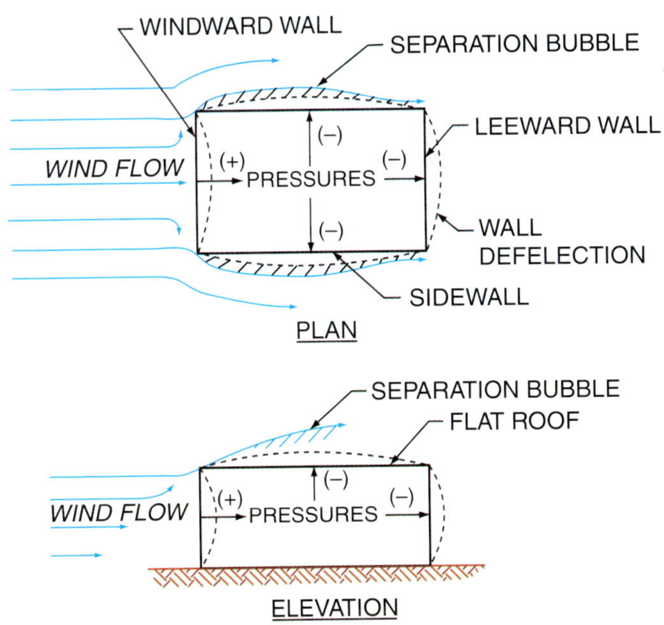

FIGURE 15-5 Wind pressures caused by external wind flow

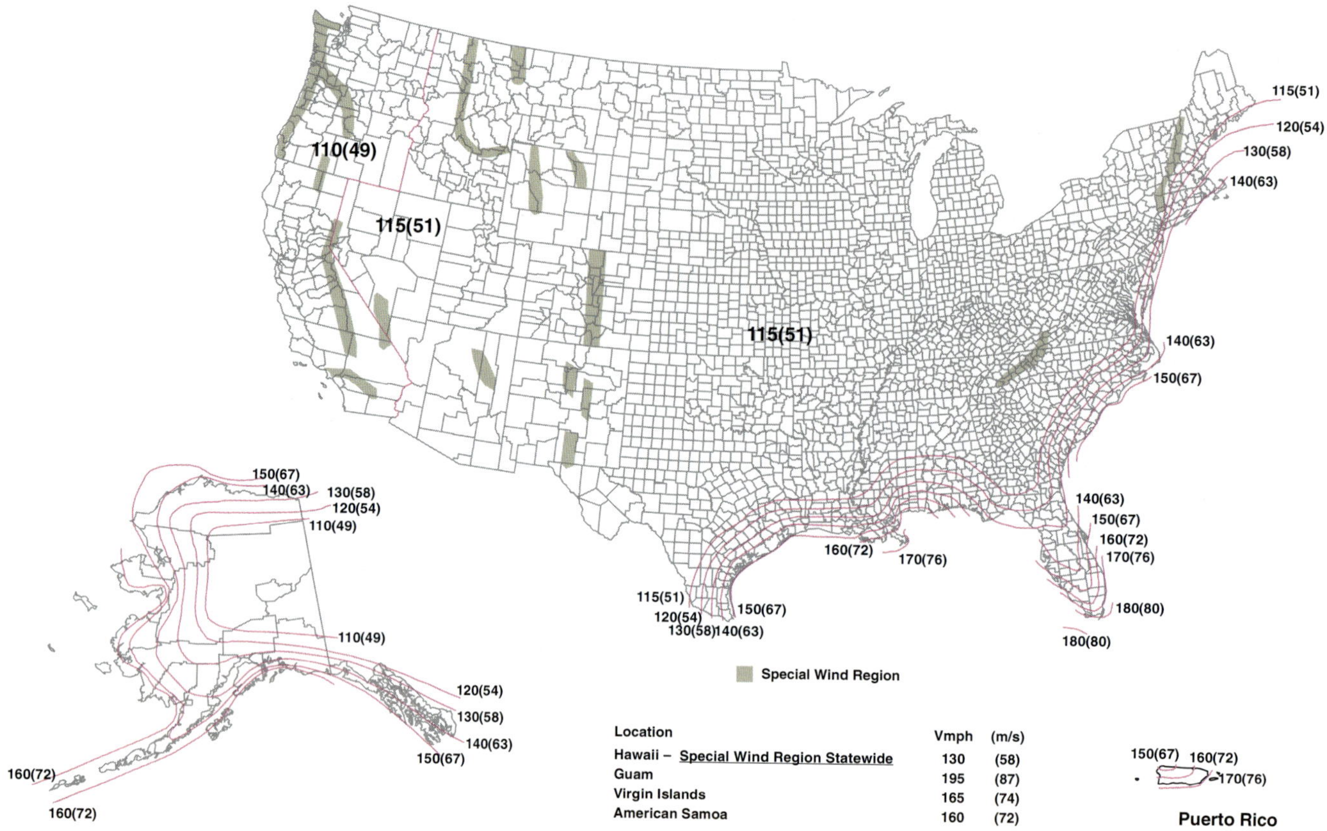

Notes:
1. Values are nominal design 3-second gust wind speeds in miles per hour (m/s) at 33 ft (10m) above ground for Exposure C category.
2. Linear interpolation between contours is permitted.
3. Islands and coastal areas outside the last contour shall use the last wind speed contour of the coastal area.
4. Mountainous terrain, gorges, ocean promontories, and special wind regions shall be examined for unusual wind conditions.
5. Wind speeds correspond to approximately a 7% probability of exceedance in 50 years (Annual Exceedance Probability = 0.00143, MRI = 700 years).

FIGURE 15-6 Ultimate design wind speeds for Risk Category II building (based on Figure 1609A)

gust measured at 33 feet above the ground. The shaded areas on the maps are special wind regions. In these areas, the local jurisdiction is responsible for determining the wind speed based on the 50-year mean recurrence interval. **[Ref. 1609.1]**

The wind speeds are used to calculate the lateral load on the exterior walls and roof and the overall stability of the building. As a building gets taller, the speed of the wind on the upper stories increases. The effect of the wind is also dependent on the topography, vegetation, and constructed features surrounding the area around the building. The code expresses these factors in categories called "surface roughness" and "exposure." For example, at areas along a coastline, the open water allows the wind to move much quicker. However, in a wooded area or urban area, the buildings and landscaping can slow the wind down. Therefore, the code requires that the design wind speed be increased in areas based on three exposure Categories, B, C, and D. **[Ref. 1609.4]**

Earthquake loads

Earthquakes can have a devastating effect on buildings (see Figure 15-7). Over the years, several earthquakes have been studied to improve the codes so the level of damage can be mitigated. An earthquake creates

FIGURE 15-7 Earthquake damage

seismic waves in the ground that shake buildings. Buildings must be designed to resist the lateral forces created by a seismic event. The IBC uses a standard map to indicate the location of low to high seismic areas in the United States. The seismic areas are based on the anticipated severity of the design earthquake ground motion at a site. **[Ref. 1613.5]**

Six seismic design categories—A, B, C, D, E, and F—are used in the design of a building. Seismic Design Category A is an area that has a low probability of an earthquake that can do substantial damage to a building. Seismic Design Category F includes areas that have the highest probability of a damaging earthquake. The most severe seismic design categories are located along the West Coast and include California, Oregon, and Washington. However, there are some areas of the Midwest and southeast part of the country that contain Seismic Design Category D areas.

The seismic design of the building is determined by ASCE 7. Many of the design requirements also come from the National Earthquake Hazards Reduction Program (NEHRP). This program was established by the U.S. Congress in 1977 to "reduce the risks of life and property from future earthquakes in the United States through the establishment and maintenance of an effective earthquake hazards reduction program." This program has been instrumental in developing earthquake design criteria in the code.

Code Basics

Surface Roughness Categories

Surface Roughness B: Urban and suburban areas, wooded areas, or other terrain with numerous closely spaced obstructions having the size of single-family dwellings or larger.

Surface Roughness C: Open terrain with scattered obstructions having heights generally less than 30 feet. This category includes flat open country, grasslands, and all water surfaces in *hurricane-prone regions*.

Surface Roughness D: Flat, unobstructed areas and water surfaces outside *hurricane-prone regions*. This category includes smooth mud flats, salt flats, and unbroken ice. ●

Flood loads

Flooding is another natural hazard that can cause substantial damage to a building. Flooding can occur anywhere in the country. However, the code concentrates on areas where the chance of flooding is more prevalent. The code identifies flood hazard areas to determine when a building must be designed to resist the loads applied by flood waters. Two different methods are used to determine flood hazard areas. The first addresses areas within a flood plain subject to a 1 percent or greater chance of flooding in any year. This is typically called the 100-year flood plain. The second method includes areas that have been designated as a flood hazard area on a community's flood hazard map or otherwise legally designated. The Federal Emergency Management Agency (FEMA) provides a Flood Insurance Rate Map (FIRM) to jurisdictions to assist in identifying these flood hazard areas. The map identifies the elevation of the 100-year base flood (see Figure 15-8). **[Ref. 1612]**

Buildings must be constructed so the lowest occupied floor is located above the base flood elevation. This requirement is intended to mitigate the amount of damage to a building caused by flooding. If the occupied area of the building is elevated above the flood, the people can reoccupy the building soon after a flood. In addition to elevating the building above the base flood elevation, the building must also be designed structurally to resist the effects of the flood waters and waves that may occur.

Unoccupied portions of a building used solely for vehicle parking, building access, or limited storage are permitted to be located below the base flood elevation. However, they must be designed so the flood

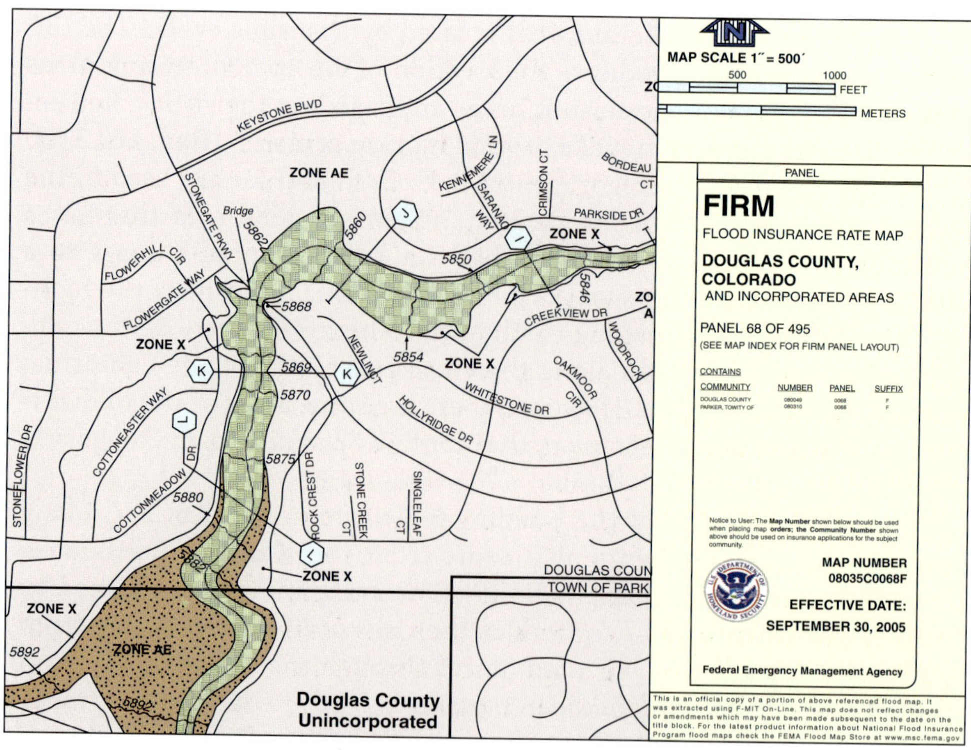

FIGURE 15-8 Sample of Flood Insurance Rate Map

waters can flow through the areas and then recede. This can be done with flood openings or by just leaving the area open so the flood waters can pass through. [Ref. 1612.5]

SOILS

Structures are ultimately supported by the ground. The condition of the soil in the ground can have a serious impact on the design of a building. There are many different types of soils. Some soils are very solid and can support a lot of weight. Some can be compacted easily and do not support much weight. There are soils that expand when they get wet and others that collapse. All of these variables must be taken into account in the structural design of the building. The soils can also exert a lateral force against a foundation, which also has to be taken into account in the design.

In order to design a structure, the design professional must know what types of soils are present on the site. The best way to determine this information is with a geotechnical investigation. The building official has the authority to require such an investigation. The investigation must be based on observation and necessary tests of the materials. This can be done by drilling into the ground and taking samples, digging test pits and taking samples, or other subsurface exploration made in appropriate locations. The studies must evaluate the soil strength and bearing capacity. They must also determine the effects of moisture on the soil-bearing capacity, compressibility, liquefaction, and expansiveness. The investigation must be conducted by a fully qualified agency and reviewed by a registered design professional. [Ref. 1803]

Excavation

The first concern regarding excavation is the location of adjacent structures. It is important that the soil next to an existing structure be supported to prevent it from collapsing out from under the building. The code requires that the existing foundation be underpinned or protected from settlement or lateral movement. The underpinning supports the soil laterally and allows the excavation to go below the foundation of the adjacent building (see Figure 15-9). [Ref. 1804.1]

It is also important that the slope of the excavation be limited so the dirt does not fall into the hole. Appendix J in the IBC addresses the slope of these surfaces. Remember, though, that the appendix chapters are only applicable when specifically adopted by the jurisdiction. Appendix J limits the slope of most cut-and-fill surfaces to no steeper than two units horizontal to one unit vertical (50 percent slope) unless a geotechnical report can justify a steeper slope. The appendix also limits the setback of the cut-and-fill slopes from lot lines to protect adjacent properties from soil erosion. [Ref. J106 and J107]

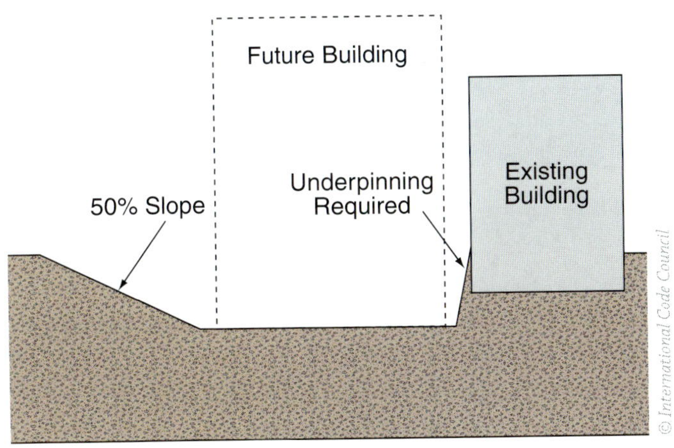

FIGURE 15-9 Excavation requirements

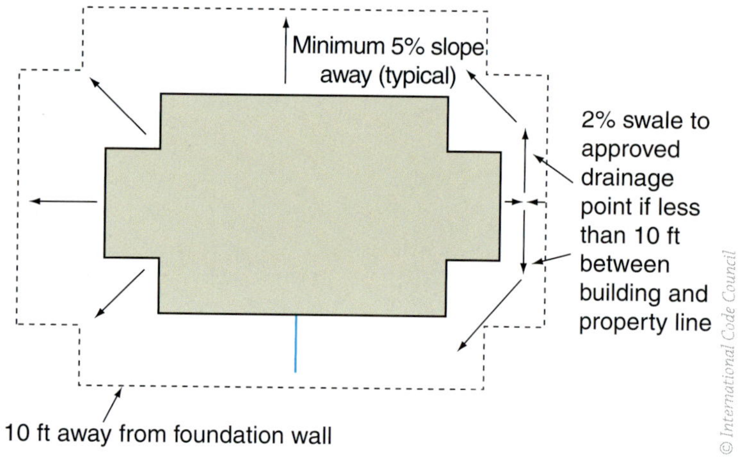

FIGURE 15-10 Minimum grading requirements

The placement of backfill against a foundation applies a large lateral force against the wall. Therefore, the IBC requires that the backfill be placed in such a manner that it does not damage the foundation. The backfill material must be free of any organic material, construction debris, cobbles, and boulders. The foundation walls should have sufficient strength to support the backfill or be provided with bracing, or the floor system must be installed on top of the wall prior to backfilling. The backfill material must be placed in lifts and compacted so it will not settle after the building is completed. **[Ref. 1804.2]**

Grading

To prevent water from ponding against a foundation wall, the IBC requires positive drainage away from the building. The ground must slope a minimum of one unit vertical in twenty units horizontal (5 percent slope) for a minimum of 10 feet perpendicular away from the face of the foundation wall. If the slope cannot be reached due to physical obstructions within the 10-foot distance, a 5 percent slope must be provided to an approved alternative method. The alternative must divert the water away from the foundation. If a swale is used as the alternative method, it must be sloped a minimum of 2 percent when located within 10 feet of the foundation. If concrete, asphalt, or other impervious materials are placed within 10 feet of the foundation, they must also be sloped a minimum 2 percent away from the building (see Figure 15-10). **[Ref. 1804.3]**

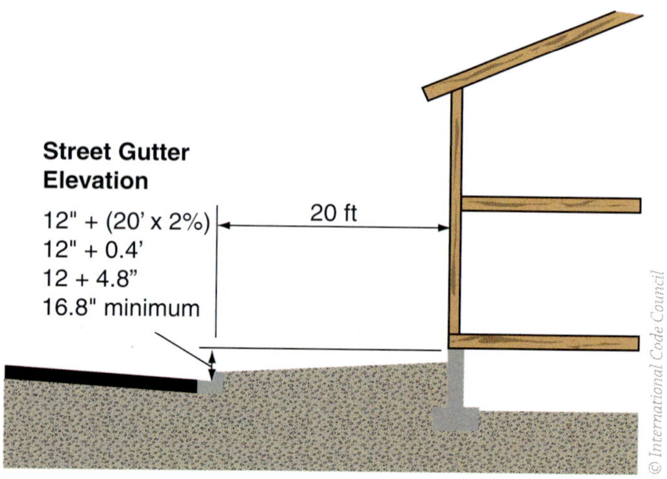

FIGURE 15-11 Foundation elevation at graded site

In addition to the slope requirements, buildings must be elevated so the exterior foundation extends above the elevation of the street gutter at a point of discharge or the inlet of an approved drainage device a minimum of 12 inches plus 2 percent. If this cannot be done, the building official has the authority to approve alternative elevations. It must be demonstrated that drainage is provided from all portions of the site away from the building and to an approved point of discharge (see Figure 15-11). **[Ref. 1808.7.4]**

CHAPTER

16 Materials

All kinds of materials are used in the construction of a building. This chapter separates these materials into two categories: structural and nonstructural. *Structural materials* are those that are used to construct the building and serve as the skeleton of the structure. *Nonstructural materials* are typically used to cover or protect that skeleton. In some cases, the structural materials are left exposed and serve as the final finished material (see Figure 16-1).

FIGURE 16-1 Concrete columns and post-tension concrete slab form the structure of this building

STRUCTURAL MATERIALS

The purpose of this chapter is to introduce the reader to the different materials used in the building. It is not intended to be a discussion on the structural design of the building using these materials. Although prescriptive requirements in the code will be discussed, additional engineering may be needed in the design of a structure. Not all structural materials are discussed in this section. Only the most common materials, such as concrete, masonry, steel, and wood, are covered.

Concrete construction

Concrete is a mixture of portland cement, aggregate, and water. In some cases, other admixtures can be added to increase or decrease the curing time, strength, and other properties of the final material. It is placed in forms in a wet condition and then cures to create a strong structural member of a building. Concrete is used throughout buildings. The most common portion is the foundation of the building, although it can be used in walls, floors, columns, beams, and even roofs. Concrete can also be used as a nonstructural material such as a non-load-bearing slab on grade.

National standards

The IBC references the American Concrete Institute's (ACI) Standard 318, *Building Code Requirements for Structural Concrete,* for the design and construction of structural concrete (see Figure 16-2). Chapter 19 of the IBC then amends portions of the standard and adds more detailed requirements not covered in the standard. The standard is intended to cover all building of the usual types for both large and small projects. **[Ref. 1905]**

Durability requirements

Even though concrete is a hard and strong material, there are conditions and elements that can affect the concrete. The IBC mitigates these conditions by requiring a minimum compressive strength based on the exposure of the concrete. The exposures are based on:

1. exposure to freezing and thawing in a moist condition or deicer chemicals;
2. exposure to sulfates in water or soil;
3. exposure to water where the concrete is intended to have low permeability; and
4. exposure to chlorides from deicing chemicals, salt, saltwater, brackish water, seawater, or spray from these sources, when the concrete has steel reinforcement.

The United States has been divided into three different levels of exposure: negligible, moderate, and severe (see Figure 16-3). Table 16-1 shows the minimum compressive strength of concrete based on these exposures.

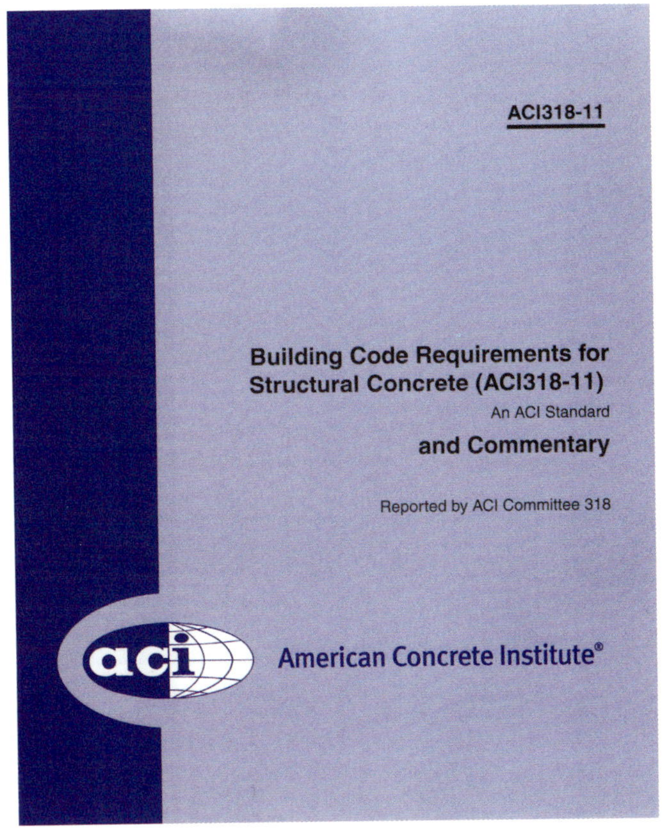

FIGURE 16-2 ACI Standard 318 is referenced by the IBC for concrete construction

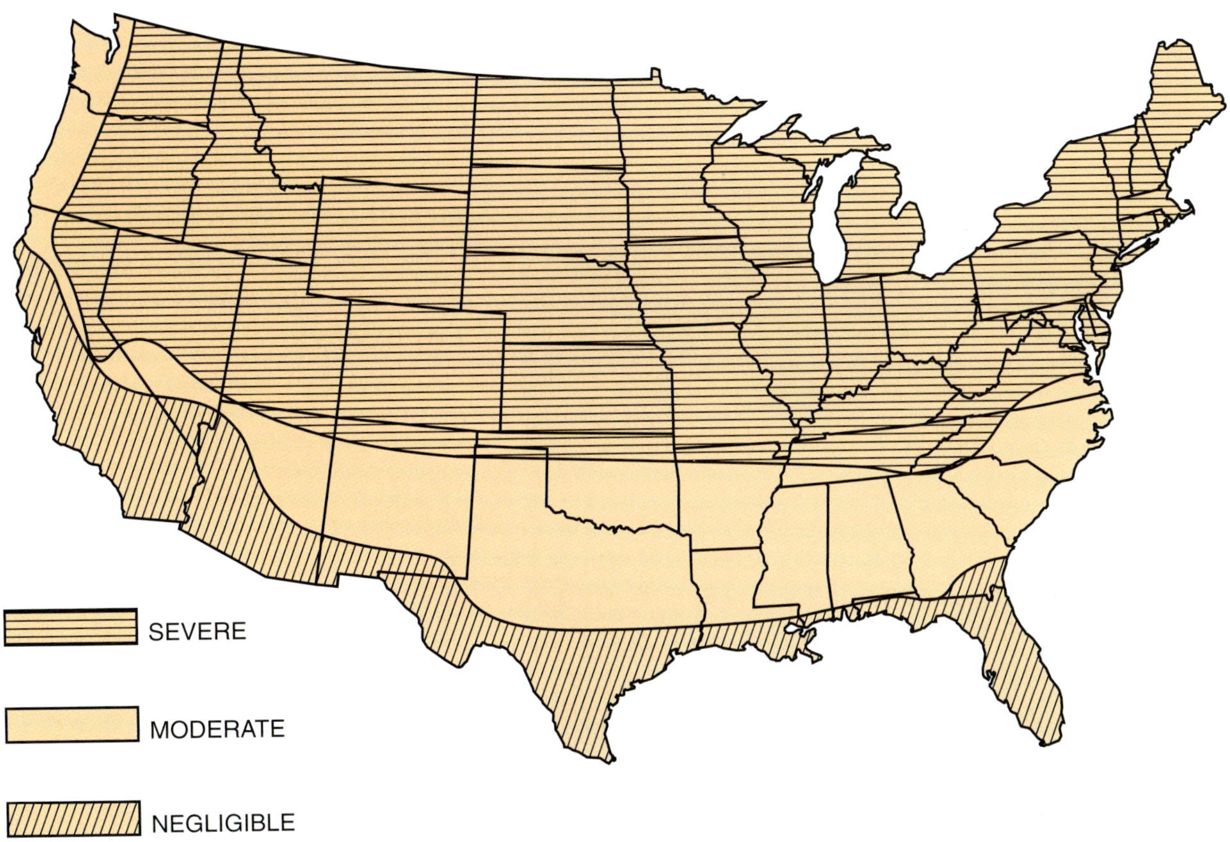

FIGURE 16-3 Exposure categories for concrete (based on IBC Figure 1904.3)

TABLE 16-1 Minimum specified compressive strength (Based on IBC Table 1904.2)

Type or Location of Concrete Construction	Minimum Specified Compressive Strength (F'_c at 28 Days, PSI)		
	Negligible Exposure	**Moderate Exposure**	**Severe Exposure**
Basement walls and foundations not exposed to the weather	2,500	2,500	2,500
Basement slabs and interior slabs on grade, except garage floor slabs	2,500	2,500	2,500
Basement walls, foundation walls, exterior walls, and other vertical concrete surfaces exposed to the weather	2,500	3,000	3,000
Driveways, curbs, walks, patios, porches, carport slabs, steps, and other flatwork exposed to the weather and garage floor slabs.	2,500	3,000	3,500

Foundations

The IBC allows concrete foundations to be either prescriptively designed or engineered. Buildings located in Seismic Design Categories C, D, E, and F must generally be designed by a design professional. In order for a foundation to be designed prescriptively, the wall must be supported laterally at the top and bottom. This means that the wall must be held in place to withstand the lateral pressure, such as soils loads applied to wall. The thickness of the wall is based on the lateral soil load, which is based on the type of soil that is present around the foundation. Some soils exert a higher lateral load than others. Therefore, the thickness of the concrete foundation wall and the amount of steel reinforcement required are based on this lateral soil load. Soils that have a very high lateral load require an engineered design.

FIGURE 16-4 Concrete forms ready for concrete placement

Formwork

Footings and foundations are created by placing wet concrete into forms that are constructed in the shape of the final configuration of the member. The formwork must be designed, fabricated, and erected in accordance with Section 6.1 of ACI Standard 318. The section requires the forms to be substantially and sufficiently tight to prevent leakage of the wet concrete as it is being placed. The forms must also be properly braced to hold them in place while the concrete is being placed (see Figure 16-4). **[Ref. 1906.1]**

The forms must remain in place until the concrete has cured enough to remain in place without the forms. The concrete must also have enough strength that it is not damaged by the removal operation. If the concrete is used in a horizontal assembly, it must be strong enough to support itself. In most cases, the contractor will shore up the concrete floor or roof system to support the assembly until the concrete reaches its specified strength. The strength is determined by casting small concrete cylinders and testing them at a testing laboratory. This process is discussed later in this chapter. If the floor or roof assembly is not provided with shoring and support, no construction loads can be placed on the assembly. **[Ref. 1906.1]**

Conduits, pipes, and sleeves are often placed in concrete. This can create a weak spot in the concrete and therefore must be taken into consideration in the design. The type of material embedded in the concrete must also be evaluated to ensure that it does not react negatively with the concrete. For example, aluminum is not permitted to be embedded in concrete. It can react to the concrete and, in some cases, to the steel reinforcement. ACI Standard 318 specifies the amount of embedded material permitted in concrete members. **[Ref. 1906.1]**

Construction joints are provided in concrete to allow the material to expand and contract due to temperature differentials. The IBC requires the construction joints to comply with the provisions of Section 6.4 of ACI Standard 318. The standard requires that the construction joints be specified on the construction documents. They must be located so they do not impair the strength of the structure and so that both gravity and lateral loads are transferred through the construction joints. **[Ref. 1906.1]**

Reinforcement

The purpose of the steel reinforcement in concrete is to resist any tension, or pulling apart, when a wall is subjected to a lateral load. The placement of reinforcement in the forms is also regulated by ACI Standard 318. Steel reinforcement is designated as a number size. The number translates to ⅛ of an inch in diameter. Therefore, a No. 4 bar is 4/8 or ½ inch in diameter and a No. 5 bar is ⅝ inch in diameter. Foundations may require both vertical and horizontal reinforcement (rebar) inside the concrete (see Figure 16-5).

Rebar must be placed inside the form in its final location prior to the placement of the concrete. It must be supported so that it does not move. ACI Standard 318 requires a minimum amount of concrete to cover the steel. For prescriptively designed foundations, a minimum of ¾-inch cover is required on the inside of a wall. The amount of concrete cover located along the outside face of a wall must be 1½ inches for No. 5 bars and smaller. For bars larger than No. 5, the minimum cover is 2 inches. ACI Standard 318 also contains minimum cover requirements for rebar (see Table 16-2). **[Ref. 1807.1.6.2]**

ACI Standard 318 also regulates the amount a piece of reinforcement is allowed to be bent to maintain its strength. Standard bends are measured as the inside radius of the bend. The minimum bending

FIGURE 16-5 Steel reinforcement used in cast-in-place concrete slab

TABLE 16-2 Concrete cover for cast-in-place concrete (nonprestressed) (Based on ACI Standard 318 Section 7.7.1)

Condition	Concrete Cover (inches)
Concrete cast against and permanently exposed to earth	3
Concrete exposed to earth or weather	
No. 6 through No. 18 bars	2
No. 5 bar, W31 and D31 wire, and smaller	1½
Concrete not exposed to weather or in contact with ground	
No. 14 and No. 18 bars	1½
No. 11 bars and smaller	¾
Beams, columns	
Primary reinforcement, ties, stirrups, spirals	1½

radius is based on the diameter of the bar. For example, the minimum bend diameter for a No. 4 rebar going around a corner of a wall is 6 d_b (6 bar diameters). So, the minimum bending radius must be 6 × ½ inch or 3 inches. Other minimum radius criteria are specified for other types of bends including stirrups, ties, and hooks. **[Ref. 1906.1]**

The rebar must be clean and free of rust and scale when it is placed in the footing or foundation forms. If it is not clean, the concrete will not adhere to the steel and form a strong structure. ACI Standard 318 requires the steel to be free from mud, oil, and other nonmetallic coatings that decrease the bond strength (see Figure 16-6). **[Ref. 1906.1]**

Quality control

The quality of the concrete has a direct relationship to its strength. The amount of water and chemicals added to the mixture can affect the strength. Therefore, the IBC requires that concrete be inspected by a special inspection agency to ensure the quality of the materials. Special inspections were briefly discussed in Chapter 2. This section discusses the specific requirements for special inspection of concrete construction. The code requires special inspection of concrete with the exception of five conditions:

FIGURE 16-6 Rebar covered with mud must be cleaned prior to placing concrete

1. Isolated spread concrete footings of buildings three stories or less above grade plane that are fully supported on earth or rock
2. Continuous concrete footings supporting walls of buildings three stories or less above grade plane that are fully supported on earth or rock where:
   a. The footings support walls of light-frame construction;
   b. The footings are designed in accordance with the prescriptive provisions of IBC Chapter 18 for shallow foundations; or
   c. The structural design of the footing is based on a specified compressive strength no greater than 2,500 pounds per square inch (psi), regardless of the compressive strength specified in the construction documents or used in the footing construction.
3. Nonstructural concrete slabs supported directly on the ground, including prestressed slabs on grade, where the effective prestress in the concrete is less than 150 psi
4. Concrete foundation walls constructed in accordance with the prescriptive provisions of IBC Chapter 18 for regular foundations
5. Concrete patios, driveways, and sidewalks, on grade

The special inspections for many materials are divided into two categories: continuous and periodic. A *continuous inspection* is the full-time observation of the work that is being done. *Periodic inspections* are part-time or intermittent observations of the work. Periodic inspections can occur during or after the work is being performed as well as after the work is completed. Concrete construction requires both continuous and periodic special inspections (see Table 16-3). **[Ref. 1705.3]**

The most common special inspection is the sampling of the concrete. Concrete must be tested every day that it is being placed, but not less than once for every 150 cubic yards (yd^3) of concrete placed nor not less than once for each 5,000 square feet of slab or wall surface. The concrete is sampled as described in ASTM Standard C172, Practice

TABLE 16-3 Required special inspections for concrete construction (Based on IBC Table 1704.4)

Continuous	Periodic
Inspection of bolts to be installed in concrete prior to and during placement of concrete where allowable loads have been increased or where strength design is used. At the time fresh concrete is sampled to fabricate specimens for strength tests, perform slump and air content tests, and determine the temperature of the concrete. Inspection of concrete and shotcrete placement for proper application techniques. Inspection of prestressed concrete: a. Application of prestressing forces. b. Grouting of bonded prestressing tendons in the seismic-force-resisting system.	Inspection of reinforcing steel, including prestressing tendons, and placement. Inspection of anchors installed in hardened concrete. Verifying use of required design mix. Inspection for maintenance of specified curing temperature and techniques. Erection of precast concrete members. Verification of in-situ concrete strength, prior to stressing of tendons in posttensioned concrete and prior to removal of shores and forms from beams and structural slabs. Inspect formwork for shape, location, and dimensions of the concrete member being formed.

FIGURE 16-7 Concrete cylinder under compression test

for Sampling Freshly Mixed Concrete. The samples are tested for consistency and the amount of air that has been entrained into the concrete. Test cylinders are molded with the concrete and cured in accordance with ASTM Standard 31, Practice for Making and Curing Concrete Test Specimens in the Field. The cylinders are typically 6 inches in diameter and 12 inches in height. After the concrete is cured, the cylinders are broken in a hydraulic test apparatus to determine the concrete's compressive strength in pounds per square inch (psi). The compressive tests are typically done at seven, fourteen, and twenty-eight days. The twenty-eight-day strength is what is specified in the IBC and ACI Standard 318 (see Figure 16-7). **[Ref. 1704.5]**

The special inspector also observes the placement of the concrete in the forms. The concrete must be conveyed from the mixer to the final location without it separating or losing materials. It must remain plastic and be continuously deposited in the forms. The concrete must then cure at a temperature above 50°F and in a moist condition for at least seven days. In cold climates, the concrete must be protected from freezing by covering the concrete in insulated concrete blankets or heating the concrete. In hot climates, the concrete must be kept cool and moist to prevent excessive temperatures or rapid water evaporation. **[Ref. 1905]**

Masonry construction

FIGURE 16-8 Concrete masonry units for construction of walls

Masonry construction has been used in buildings for centuries (see Figure 16-8). It is used structurally in foundations, bearing walls, and columns, as well as in nonstructural applications like veneers. The use of masonry in veneers was discussed in Chapter 13. This section discusses the use of masonry as a structural material in buildings. The six different design methods used in masonry construction are allowable stress design, strength design, prestressed masonry, empirical design, glass unit masonry, and masonry veneer. Each of these design methods is covered in several standards published by the Masonry Standards Joint Committee (MSJC), the American Concrete Institute (ACI), the Structural Engineering Institute of the American Society of Civil Engineers (ASCE), and The Masonry Society (TMS). Though each of these organizations has developed standards for different types of masonry construction, this section does not address the specifics of these standards.

Foundations

As with concrete foundations, the IBC provides a prescriptive method for designing masonry foundations. The thickness of the walls is determined by the lateral soil load against the foundation. Steel reinforcement is also required in masonry foundations. In some cases, the cells of the hollow or solid masonry units must be filled with grout. Masonry foundations for buildings located in the higher seismic design categories must be provided with additional engineering to ensure the stability of the masonry during an earthquake. **[Ref. 1807.1.6.3]**

Materials

As already noted, there are many different types of masonry materials. Masonry units can be manufactured with a multitude of raw material. Each of the units has a different purpose and different advantages. Some types of masonry provide a better fire-resistance rating, whereas others provide a stronger structural strength. The IBC regulates many of these materials by referencing the standards discussed before.

Four major types of masonry units are regulated by the IBC: concrete, clay or shale, stone, and glass masonry units. Glass masonry units are not discussed in this book. Concrete masonry is a building unit or block that is larger in size than 12 inches by 4 inches by 4 inches and made of cement and suitable aggregates. Clay masonry is a building unit larger than a brick and is composed of clay, shale, fired clay, or a mixture of these. Stone masonry is composed of field, quarried, or cast stone units. All of these are bonded together with mortar (see Figure 16-9). **[Ref. 202]**

FIGURE 16-9 Clay brick veneer placed in front of concrete masonry wall

Mortar is a mixture of cement, fine aggregates, and water that is used to bond masonry units. Four different types of mortar are used in masonry construction: Type M, Type S, Type N, and Type O. Type M mortar has the highest compressive and tensile bond strengths. Type S mortar is a general-purpose mortar with high compressive and tensile bond strengths. Type N mortar is a general-purpose mortar with medium compressive and tensile bond strengths. Type O is a low-strength mortar suitable for use in non-load-bearing construction and in construction in which the masonry will not be subject to severe weathering or freeze–thaw cycling. The materials used in mortar are also required to conform to ASTM and TMS standards. **[Ref. 2103.8]**

Grout is commonly used to fill the cells of hollow masonry units and secure vertical steel reinforcement in the cells. Two types of grout are used in reinforced masonry: fine and coarse. *Fine grout* is comprised of cement, fine aggregate, and water. *Coarse grout* includes the same material as fine grout, but pea gravel limited to ⅜ inch in diameter is added to the mix. Steel reinforcing must comply with the same requirements as it does for concrete structures. It must be the proper size and be free of any foreign materials to ensure a proper bond between the grout and the steel. **[Ref. 2103.12]**

Quality control

Special inspections are required on masonry construction similar to concrete. The level of special inspections on masonry construction is based on the risk categories discussed in Chapter 15. As a review, Category IV risk buildings are structures that are very important to a community during a disaster. Category I buildings are those that are the least important to the community during a disaster. Empirically designed masonry, glass unit masonry, and masonry veneer in Category IV occupancies and engineered masonry in Category I, II, or III risk categories must be provided with a higher level of inspections than engineered masonry in Category IV occupancies. Both continuous and periodic inspections are required for masonry construction at both levels of special inspection. Continuous inspections include verification of the grout consistency, welding of reinforcing bars, and the grout placement. Periodic inspections include location and size of reinforcement, construction of mortar joints, grout cell cleanliness, and the protection of the masonry during hot or cold weather. **[Ref. 1705.4]**

The testing of masonry construction is also required by the IBC. Two methods of testing are used to ensure that the quality of the material being used complies with the structural design: the unit strength method and the prism test method. The use of the *unit strength method* is determined based on the strength of the unit used and the type of mortar. Conditions set on using this method include the thickness of the sampling and testing of the masonry unit, the thickness of the mortar joint, and the strength of the grout. The *prism test method* involves the construction of a sample wall constructed in the same way as it is built in the field. A minimum of three prisms are made and tested by a special inspection agency, similar to the test used to determine the compressive strength of concrete (see Figure 16-10). The IBC provides an alternative to making the prisms by allowing the use of prisms cut from the masonry construction, but this is not the preferred method.

FIGURE 16-10 Concrete masonry unit prisms ready for testing

Steel construction

Structural steel under the IBC is a member consisting of rolled steel structural shape (see Figure 16-11). Many different structural shapes are used in the design and construction of a building. Steel joists are also common structural steel elements of a building. This section discusses the IBC requirements for these types of members. The majority of the requirements for steel design are based in the standards promulgated by the American Institute of Steel Construction (AISC) and the Steel Joist Institute (SJI). In fact, the IBC specifically requires the design, fabrication, and erection of structural steel used for buildings and structures to be done in accordance with AISC Standard 360, Specification for Structural Steel Buildings. Steel joist designs and installations are required to comply with four different SJI standards, depending on the type of joist being used. The IBC then adds additional requirements for buildings located in high seismic design categories based on the AISC standards. Unlike concrete and masonry, the requirements for steel construction are left to the standards. Very few additional requirements are addressed in the IBC. **[Ref. 2205 and 2206]**

FIGURE 16-11 Steel observation tower at Kennedy Space Center

Protection

Chapter 8 briefly discussed the protection of steel from the effects of fire. This section discusses the protection of steel from the effects of the natural elements. Rust can have an injurious effect on steel. Therefore, the IBC requires the steel to be protected with an approved coating to protect the steel from corrosion. The painting of structural steel members must comply with AISC Standard 360. Steel joists must also be protected in accordance with the applicable SJI standard. When steel is erected, the protection of the steel must be maintained. If the paint is scratched or chipped during erection, an approved paint or primer should be applied. However, if the steel is going to be protected with sprayed applied fireproofing, the steel must be either bare steel or painted with an approved primer. The steel is often left unprimed to ensure a better adhesion of the fireproofing. **[Ref. 2203.2]**

Connections

Two methods are used to connect structural steel members: welding and bolting. In both cases, special inspections are required to be provided by an approved agency. The welding of connections must be inspected on a continuous and periodic basis depending on the size and importance of the welds. Single-pass fillet welds of 5⁄16 inch or less and welding of floor and roof decking are permitted to be periodically inspected. Other welds, including complete and partial joint penetration groove welds, multipass fillet welds, single-pass fillet welds larger than 5⁄16 inch, and plug and slot welds, must be continuously inspected.

Two types of joints are used in bolted connections. A *slip-critical joint* is a bolted connection designed to resist movement by friction on the contact surface of the connection under the clamping forces of the bolts. A *pretensioned joint* is a joint with high-strength bolts tightened to the specified minimum pretension. High-strength bolting is inspected periodically except for pretensioned and slip-critical joints using the turn of the nut without matchmarking or calibrated wrench methods of installation. These types of connections are rare in today's construction methods. Most bolting connections are either snug-tight joints or twist-off bolts. In the case of *snug-tight joints,* the nuts are just tightened with a few impacts of an impact wrench or the full effort of an ironworker using an ordinary spud wrench to bring the connected plies into firm contact. A *twist-off bolt* is designed to break off the end of the bolt that is inserted into an impact wrench at a specified torque. These bolts are relatively easy to inspect. The inspector needs to identify and confirm the type of bolt and that the end has been sheared off (see Figure 16-12).

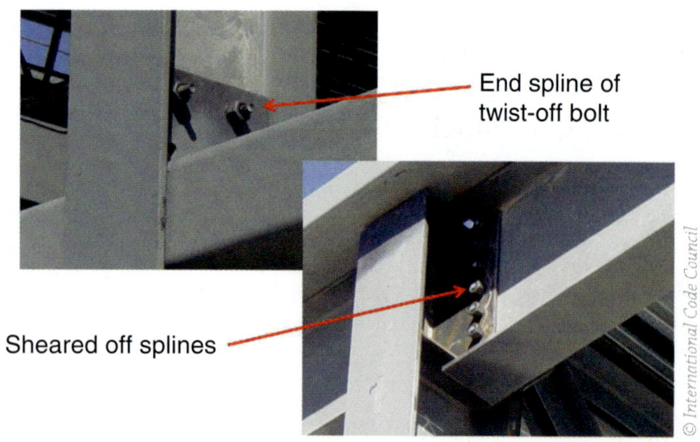

FIGURE 16-12 Twist-off bolt connection

Quality control

In the previous section, we talked about the special inspection requirements for the connections. In addition to these inspections, the IBC requires periodic inspections of the steel frame to verify compliance with the details shown on the approved construction documents, such as bracing, stiffening, member locations, and proper application of joint details at each connection. The inspections also must verify that the structural steel and decking are the materials specified on the construction documents.

The IBC also requires special inspection for the fabrication of steel joists. A fabricator must either have the fabrication of the joists inspected by an inspection agency or become an approved fabricator. An approved fabricator must have a written procedural and quality control manual and periodic auditing of fabrication practices by an approved special inspection agency. When the fabrication is completed, the approved fabricator must submit a certificate of compliance to the building official stating that the work was performed in accordance with the approved construction documents.

FIGURE 16-13 One-story wood frame construction

Wood

The use of wood as a structural element of a building has been a tradition for many years (see Figure 16-13). The two types of wood construction in buildings are conventional light-frame construction and heavy timber construction. Conventional light-frame construction is a type of construction in which the primary structural elements are created by a system of repetitive wood framing members. Heavy timber construction was briefly discussed in Chapter 4. It utilizes large-dimensioned lumber as the structural elements. The IBC regulates the minimum size of the structural members in heavy timber construction. This section is limited to the discussion of conventional light-frame construction. Conventional light-frame construction is limited to smaller buildings with limited loading requirements and restrictions on roof spans and the types of uses permitted. [Ref. 2308.2]

Protection

Because wood is a natural material, it is susceptible to damage from moisture and insects. The IBC provides requirements that are intended to mitigate that damage. The code requires either a species of wood that is naturally durable or preservative-treated lumber. Naturally durable wood is a specific species of wood that is decay and termite resistant. Preservative-treated lumber is wood that has been vacuum-pressure impregnated with a chemical that reduces the susceptibility to damage by fungi, insects, or marine borers. There are seven locations where naturally durable wood or preservative-treated lumber is required above ground.

Wood joists and the bottom of wood structural floors without joists, which are closer than 18 inches to the exposed ground, or wood girders that are closer than 12 inches to the exposed ground in crawl spaces or unexcavated areas located within the perimeter of the building

> **Code Basics**
>
> **Naturally Durable Wood**
>
> Decay Resistant:
> Redwood, cedar, black locust, and black walnut
>
> Termite Resistant:
> Redwood, Alaska yellow-cedar, Eastern red cedar, and both heartwood and all sapwood of Western red cedar

foundation must be of naturally durable or preservative-treated wood. The requirement applies to the entire floor construction including posts, girders, joists, and subfloor. As long as the framing members are held above these dimensions, any type of wood species can be used.

When the clearance between wood siding and the exposed earth is less than 6 inches or less than 2 inches from the surface of concrete steps, porch slabs, patio slabs, and similar surfaces, the siding must be naturally durable or preservative-treated wood. If the wood siding is located higher than these two dimensions, any species of wood can be used. Naturally durable or preservative-treated wood is also required for any wood framing members including exterior sheathing that is resting on foundation walls that are less than 8 inches above the exposed earth (see Figure 16-14) or any furring strips attached directly to the interior of masonry or concrete walls located below grade.

Posts and columns supporting permanent structures that are supported by concrete or masonry slab or footing that is in direct contact with the earth must be naturally durable or preservative-treated wood. However, there are two conditions for which that any type of wood can be used. The first is when wood posts or columns are exposed to the weather or located in basements or cellars and they are supported by metal pedestals projecting 1 inch above the slab or deck and 6 inches above exposed earth (see Figure 16-15). The second condition is when the posts or columns are located in an enclosed crawl space or unexcavated areas located within the periphery of the building and are supported by a concrete or metal pedestal at a height greater than 8 inches from the exposed ground. In both cases, the wood must also be separated from the soil by an impervious moisture barrier.

FIGURE 16-14 Bottom plate must be treated lumber or kept 8 inches above grade

Wood in direct contact with the earth or immersed in freshwater must also be naturally durable wood or preservative-treated wood for those purposes. If a column or post is embedded in concrete that is exposed to the weather or in direct contact with the earth, it can only be preservative-treated wood. When wood structural members are used to support moisture-permeable floors or roofs and they are exposed to the weather (such as concrete or masonry slabs), naturally durable or preservative-treated wood must be used. If the wood structural members are separated from the slab with an impervious moisture barrier, any type of lumber can be used.

FIGURE 16-15 Wood post supported by metal pedestal projecting 1 inch above concrete

Grade marks

To assist inspectors with identifying the species of wood used in a structure, lumber must be identified by the grade mark of a lumber grading or inspection agency. The agency must be approved by an accreditation body that complies with the U.S. Department of Commerce (DOC) PS20 American Softwood Lumber Standard, or equivalent. This standard sets rules for the grading practices and identification of the lumber. As the wood is milled, grading inspectors visually grade the lumber. A stamp is placed on the lumber showing the species or species grouping, grade, and moisture content at time of surfacing; the grading agency; and the mill name or grader's number (see Figure 16-16). **[Ref. 2303.1.1]**

FIGURE 16-16 Lumber grade mark

Wood structural panels used for sheathing on floors, walls, roofs, siding, diaphragms, and built-up members must also be identified for grade and glue type by the trademarks of an approved testing agency (see Figure 16-17). The panels must comply with DOC Standard PS 1 or Standard PS 2 depending on the type of panel being manufactured. In addition, wood structural panels that are permanently exposed in outdoor applications shall be of exterior type. Wood structural panels used in roof sheathing that are exposed to the outdoors on the underside are permitted to be interior-type bonded panels with exterior glue, Exposure 1. Exposure 1 panels are permitted to be used in applications where construction delays may occur before any protection is provided. It is a bond classification determined by APA — The Engineered Wood Association. **[Ref. 2303.1.4]**

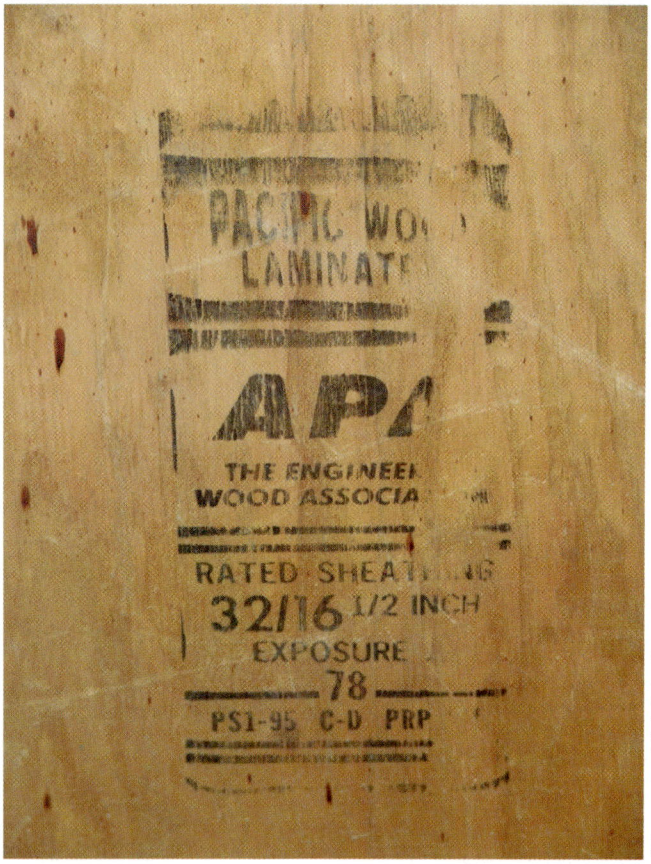

FIGURE 16-17 Wood structural panel marking

Preservative-treated lumber must be provided with a quality mark providing information from an inspection agency that maintains continuing supervision, testing, and inspection over the quality of the treated lumber. The lumber must comply with the American Wood Protection Association (AWPA) standards. The quality mark must include the following information:

1. Identification of treating manufacturer
2. Type of preservative used
3. Minimum preservative retention (pcf)
4. End use for which the product is treated
5. AWPA standard to which the product was treated
6. Identity of the accredited inspection agency

Cutting, boring, and notching

A wood framing member's strength is dependent on the fibers in the wood holding together under load. The IBC therefore regulates the amount of notching and drilling of joists, rafters, and studs.

A notch on the end of a floor joist, ceiling joist, or rafter is not permitted to exceed ¼ the depth of the member. A notch in the top or bottom of a joist or rafter cannot exceed 1/6 the depth of the joist or rafter and cannot be located in the middle third of the span. Holes bored in a joist or rafter are not permitted to be located within 2 inches of the top or bottom of the member. The diameter of a hole drilled in a joist or rafter cannot exceed ⅓ the depth of the member (see Figure 16-18). **[Ref. 2308.8.2 and 2308.10.4.2]**

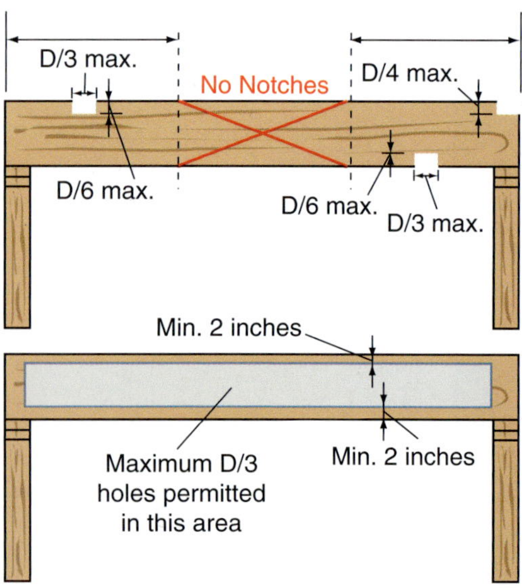

FIGURE 16-18 Joist/rafter notching and holes

The amount and size of notches and bored holes are also limited in wall studs. A stud installed in an exterior bearing wall or bearing partition is permitted to be cut or notched to a depth not exceeding 25 percent of the width of the stud. Holes in exterior and bearing walls are limited to 40 percent of the width of the stud. If the studs are doubled, the hole can be up to 60 percent of the stud width as long as no more than two successive doubled studs are bored. Nonbearing partitions that do not support loads other than the weight of the partition are permitted to be cut and notched up to 40 percent of the stud's width. Holes in nonbearing partitions are permitted to be no larger than 60 percent of the width of the stud (see Figure 16-19). **[Ref. 2308.9.10 and 2308.9.11]**

FIGURE 16-19 **Plumbing vent lines pass through holes bored in nonbearing wall studs**

Floor construction

The IBC provides a prescriptive design for wood floor systems in conventional light-frame construction. The code provides basic span limitations for the most common wood species used in these types of buildings. For spans of materials not specified in the code, the *Span Tables for Joists and Rafters* published by the American Forest & Paper Association (AF&PA) can be used. The floor joist span tables in the IBC can be used for live loads of 30 and 40 pounds per square foot (psf) and dead loads of 10 and 20 psf. Joist spacing of 12″, 16″, 19.2″, and 24″ on center are included in the tables. **[Ref. 2308.8]**

Floor joists must be supported at each end with a minimum amount of bearing. The minimum bearing is 1½ inches on wood and metal and not less than 3 inches on masonry. The IBC also permits a joist to bear on a 1-inch-by-4-inch ribbon strip with the joist nailed to the adjacent stud. This type of framing is typically referred to as *balloon framing*. When floor joists frame from opposite sides of a beam, girder, or wall, the joists must be lapped at least 3 inches or tied together in an approved manner. If the joist frames in the side of a girder or beam, it must be supported by an approved framing anchor (joist hanger) or 2-inch-by-2-inch ledger strip (see Figure 16-20). **[Ref. 2308.8.1]**

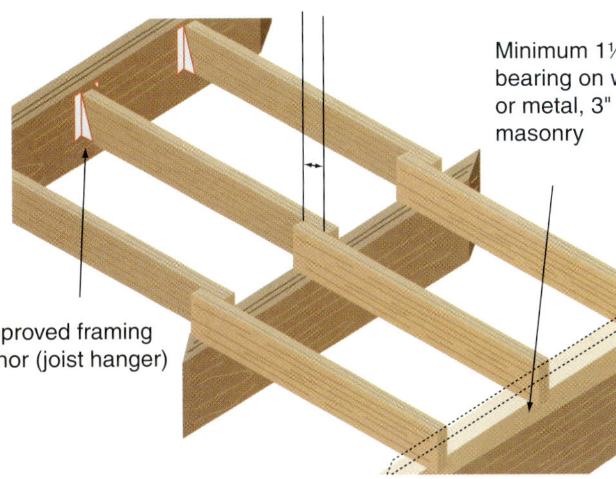

FIGURE 16-20 **Floor framing details**

When openings for stairs or shafts are place in wood-framed floors, the IBC requires specific framing details around these openings. If the header of the opening exceeds 4 feet in length, the trimmer and header must

EXAMPLE

Determine the maximum span size and maximum spacing of a floor joist of Hem Fir #2 with a span of 15 feet and supporting a 40 psf live load and 10 psf dead load using Table 16-4.

TABLE 16-4 Floor joist spans for common lumber species (Based on IBC Table 2308.8)

Joist Spacing (inches)	Species and Grade		Dead Load = 10 psf			
			2 × 6	2 × 8	2 × 10	2 × 12
			Maximum Floor Joist Spans			
			(ft – in)	(ft – in)	(ft – in)	(ft – in)
12	Douglas fir	#2	11–10	15–7	19–10	23–0
	Hem fir	#2	11–0	14–6	18–6	22–6
	Southern pine	#2	11–10	15–7	19–10	24–2
	Spruce-pine-fir	#2	11–3	14–11	19–0	23–0
16	Douglas fir	#2	10–9	14–1	17–2	19–11
	Hem fir	#2	10–0	13–2	16–10	19–8
	Southern pine	#2	10–9	14–2	18–0	21–1
	Spruce-pine-fir	#2	10–3	13–6	17–2	19–11
19.2	Douglas fir	#2	10–1	12–10	15–8	18–3
	Hem fir	#2	9–5	12–5	15–6	17–1
	Southern pine	#2	10–1	13–4	16–5	19–3
	Spruce-pine-fir	#2	9–8	12–9	15–8	18–3
24	Douglas fir	#2	9–1	11–6	14–1	16–3
	Hem fir	#2	8–9	11–4	13–10	16–1
	Southern pine	#2	9–4	12–4	14–8	17–2
	Spruce-pine-fir	#2	8–11	11–6	14–1	16–3

SOLUTION:

Joist spacing	Minimum joist size
12 in. O.C.	2 × 10
16 in. O.C.	2 × 10
19.2 in. O.C.	2 × 10
24 in. O.C.	2 × 12

be doubled or an equivalent cross section of lumber must be provided. If the header length exceeds 6 feet, it must be supported by framing anchors or joist hangers unless it is resting on a beam, partition, or wall. Bearing walls supported by floor framing must be supported on a beam, girder double joists, or walls. If the bearing wall is perpendicular to the joists, it cannot be offset more than the depth of the joists from the beam, girder, or wall below. If the joists are designed to support the load of the bearing wall, the offset is not limited. **[Ref. 2308.8.3 and 2308.8.4]**

When floor joists are installed in a repetitive manner and a load is applied to the framing, the joists can roll over like dominos. Therefore, the IBC requires lateral support of the joists to prevent this. The amount of lateral support is dependent on the depth of the joist (see Table 16-5). **[Ref. 2308.8.5]**

The sheathing used for the floor structure must also be designed to support the anticipated loads. The IBC provides prescriptive tables for different types of sheathings including sheathing lumber and wood structural panel sheathing. The common sheathing used is the wood structural panels. The thickness of the panel and the panel rating are used to determine its strength. The most common panel used for floor sheathing is a ¾-inch panel, 48/24 rated. The first number in the panel rating is the maximum span for roof sheathing with all of the edges supported. The second number is the maximum span for floor sheathing with tongue-and-groove joints or with all of the edges supported. Therefore, a ¾-inch, 48/24 rated, wood structural panel is an acceptable floor sheathing as long as the floor joists are not spaced more than 24 inches on center. **[Ref. 2304.7]**

Walls

Walls used in conventional light-frame construction can also be designed prescriptively, according to the IBC. The size, height, and spacing of the studs in bearing walls are based on what they are supporting. The size of wood studs in nonbearing walls is based on the stud height and spacing (see Table 16-6). The studs must be installed with their wide dimension perpendicular to the wall and continuous from the bottom to top plate.

TABLE 16-5 Lateral support (IBC Section 2308.8.5)

Joist Depth-to-Thickness Ratio	Lateral Support Required
5:1	One edge of joist held in line for the entire span
6:1 or more	One line of bridging for each 8 feet of span; or both edges of joists held in line for the entire length

TABLE 16-6 Size, height, and spacing of wood studs (Based on IBC Table 2308.9.1)

Stud Size (inches)	Bearing Walls				Nonbearing Walls	
	Stud Height (feet)[a]	Supporting Roof and Ceiling Only	Supporting One Floor, Roof, and Ceiling	Supporting Two Floors, Roof, and Ceiling	Stud Height (feet)[a]	Spacing (inches)
		Spacing (inches)				
2 × 4	10	24	16	NP	14	24
3 × 4	10	24	24	16	14	24
2 × 5	10	24	24	NP	16	24
2 × 6	10	24	24	16	20	24

NP = Not permitted

a. Listed heights are distances between points of lateral support placed perpendicular to the plane of the wall. Increases in unsupported height are permitted where justified by an analysis.

> ## Code Basics
> **Braced Wall Panel Methods**
> 1. Nominal 1-inch-by-4-inch continuous diagonal braces let into top and bottom plates and intervening studs
> 2. Minimum ⅝-inch-thick wood boards applied diagonally on studs spaced not over 24 inches on center
> 3. Minimum ⅜-inch-thick wood structural panel sheathing for 16-inch or 24-inch stud spacing
> 4. Minimum ½-inch fiberboard sheathing panels applied vertically or horizontally on studs spaced not over 16 inches on center
> 5. Minimum ½-inch gypsum board on studs spaced not over 24 inches on center and nailed at 7 inches on center
> 6. Particleboard wall sheathing panels
> 7. Portland cement plaster on studs spaced 16 inches on center
> 8. Hardboard panel siding •

The code generally requires a minimum of three studs in each corner of an exterior wall. The top plates must be doubled in bearing and exterior walls to provide overlapping plates at corners and intersections with other partitions. The plates must be a minimum of 2 inches in nominal depth and as wide as the studs. End joints in the double top plates must be offset a minimum of 48 inches from each other. **[Ref. 2308.9.1 and 2308.9.2]**

When openings are provided in walls, headers must be provided to support the loads in bearing walls. The IBC provides prescriptive tables for designing both headers and girders because they both serve similar purposes. Wood headers are required to be a minimum of two pieces of 2-inch nominal lumber set on edge and nailed together with 16d common nails spaced 16 inches on center along the edge of the header. Solid lumber of equivalent size to the built-up header can also be used. The headers must be supported on each end by wood studs with a minimum bearing of 1½ inches for the full width of the header. Headers in non-bearing walls do not need to be designed because they do not support any loads. They are permitted to be framed using single studs and headers (see Figure 16-21).

The beginning of this chapter discussed the loads applied to buildings under different conditions. One of the loads mentioned was lateral loads applied by wind and/or seismic activity. Conventional light-frame construction requires wall bracing to mitigate the impact of these lateral loads. A braced wall panel acts as a stiffener to the wall studs and prevents the studs from racking under these loads. Braced wall lines are typically the exterior walls of the building. However, if the structure is large enough, additional braced wall lines must be provided perpendicular to the exterior walls to support the lateral load against the outside wall (see Figure 16-22). The maximum spacing of braced wall lines is 35 feet in Seismic Design Categories A, B, and C and 25 feet in D and E.

Eight different methods can be used to provide bracing for braced wall lines. The most common way in today's construction is with wood structural panel sheathing. The type and amount of braced wall panels

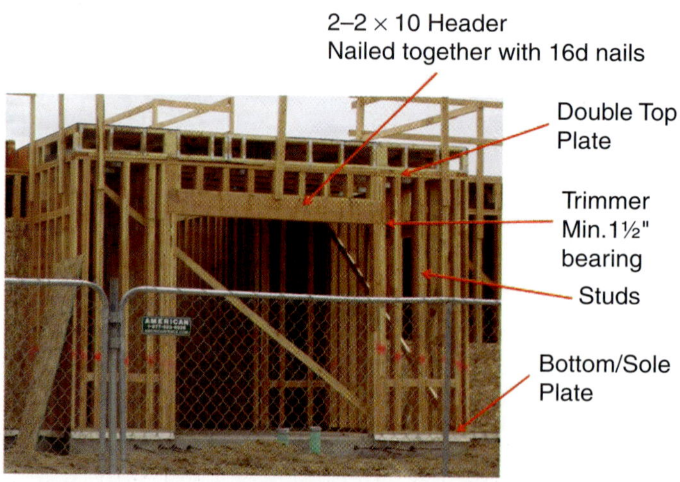

FIGURE 16-21 Typical wood stud wall framing

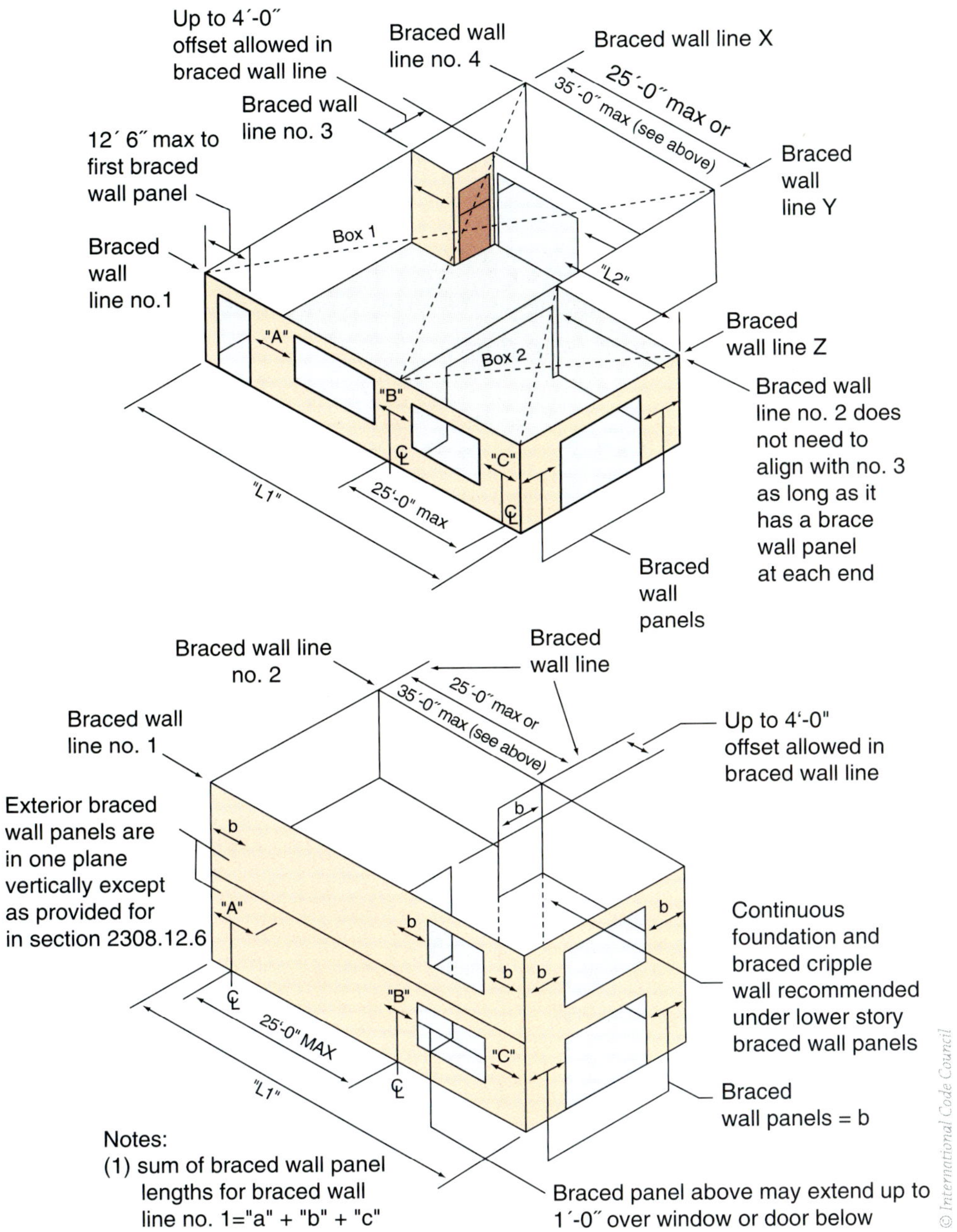

FIGURE 16-22 Basic components of the lateral bracing system

are based on the seismic design category and the location of the wall in relation to the story of the building (see Table 16-7). Remember, the conventional light-frame construction is limited to three stories. Therefore, buildings larger than this must be designed using the engineering standards specified in the IBC. Buildings located in Seismic Design Categories D and E must also be engineered.

Ceilings and roofs

The design and construction of ceilings and roofs are very similar to those for the floor systems. Similar prescriptive span tables for ceiling joists and roof rafters are provided in the IBC. Ceiling joists not only support the weight of the ceiling finishes but also provide a tie between the ends of

TABLE 16-7 Braced wall panels (Based on IBC Table 2308.9.3)

Seismic Design Category	Condition	Bracing Method								Braced Panel Location and Length
		1	2	3	4	5	6	7	8	
A and B	One-story, top of two- or three-story	X	X	X	X	X	X	X	X	Must begin within 12½ feet from each end of a braced wall line and not more than 25 feet on center.
	First story of two-story or second story of three-story	X	X	X	X	X	X	X	X	
	First story of three-story	NP	X	X	X	X	X	X	X	
C	One story of top of two-story	NP	X	X	X	X	X	X	X	Must begin within 12½ feet from each end of a braced wall line and not more than 25 feet on center.
	First story of two-story	NP	X	X	X	X	X	X	X	Must begin within 12½ feet from each end of a braced wall line and not more than 25 feet on center, but total length shall not be less than 25% of building length.

NP = Not permitted.

the roof rafters to prevent the outward thrust of the rafters at the top of the walls. The maximum spans for ceiling joists are based on whether the attic will be used for storage or not. Rafter span tables are based on either the live load or snow loads. The live load tables are used for buildings constructed in areas without snow loads. **[Ref. 2308.9.2 and 2308.9.3]**

Rafters must be framed directly opposite each other at the ridge. The ridge must be a minimum 1-inch nominal thickness at ridges. It must be the same depth as the cut end of the rafter. A 2-inch nominal thick valley or hip rafter, the same depth of the cut end of the rafter, must be provided at valleys and hips. The rafters must be tied together by either ceiling joists as noted previously or by rafter ties. They must be nailed together and to the top plate of the wall assembly they bear on. The IBC specifies the number and type of fasteners based on whether or not a snow load is present. If rafter ties are used, they must be spaced no more than 4 feet on center. The requirements for framed openings and lateral support are identical to the floor framing requirements. **[Ref. 2308.10.4]**

Roof sheathing is also similar to the requirements for floor construction. The thickness and panel rating of structural panel roof sheathing are based on whether or not the edges are supported and the load that it is supporting (see Table 16-9). The maximum span and the load must be evaluated to ensure the proper sheathing is being used. The installation of structural panel sheathing in both floors and roof must be installed in accordance with the manufacturer's installation instructions. **[Ref. 2304.7.2]**

Crawl space and attic access

Access to concealed spaces like crawl spaces and attics is needed to service equipment that may be located in these areas. The IBC requires access

EXAMPLE

Determine the minimum size and maximum spacing for a Douglas Fir #2 rafter with a span of 14 feet with a ceiling attached to the bottom of the rafters. The ground snow load is 30 psf and the dead load is 10 psf. Refer to Table 16-8.

TABLE 16-8 Rafter spans for common lumber species (Ground snow load = 30 psf, ceiling attached to rafters) (Based on IBC Table 2308.10.3(5))

Joist Spacing (inches)	Species and Grade		Dead Load = 10 psf				
			2 × 4	2 × 6	2 × 8	2 × 10	2 × 12
				Maximum Floor Joist Spans			
			(ft – in)	(ft – in)	(ft – in)	(ft – in)	(ft – in)
12	Douglas fir	#2	8–7	13–6	17–5	21–4	24–8
	Hem fir	#2	8–0	12–7	16–7	21–0	24–4
	Southern pine	#2	8–7	13–6	17–10	22–3	Note a
	Spruce-pine-fir	#2	8–3	12–11	17–0	21–4	24–8
16	Douglas fir	#2	7–10	11–11	15–1	18–5	21–5
	Hem fir	#2	7–3	11–5	14–11	18–2	21–1
	Southern pine	#2	7–10	12–3	16–2	19–3	22–7
	Spruce-pine-fir	#2	7–6	11–9	15–1	18–5	21–5
19.2	Douglas fir	#2	7–4	10–11	13–9	16–10	19–6
	Hem fir	#2	6–10	10–9	13–7	16–7	19–3
	Southern pine	#2	7–4	11–5	14–9	17–7	20–7
	Spruce-pine-fir	#2	7–0	10–11	13–9	16-10	19–6
24	Douglas fir	#2	6–8	9–9	12–4	15–1	17–6
	Hem fir	#2	6–4	9–7	12–2	14–10	17–3
	Southern pine	#2	6–10	10–2	13–2	15–9	18–5
	Spruce-pine-fir	#2	6–6	9–9	12–4	15–1	17–6

a. Span exceeds 26 feet in length. Check sources for availability of lumber in lengths greater than 20 feet.

SOLUTION:

Joist Spacing	Minimum Joist Size
12 in. O.C.	2 × 8
16 in. O.C.	2 × 8
19.2 in. O.C.	2 × 10
24 in. O.C.	2 × 10

openings for these spaces to be provided that access. Crawl spaces must be provided with a minimum 18-inch-by-24-inch access opening. Attics that have a clear height over 30 inches must have an opening that is at least 20 inches by 30 inches. There must be a minimum clear headroom height of 30 inches at or above the access opening. If mechanical equipment is installed in a crawl spaces or attic, the *International Mechanical*

TABLE 16-9 Allowable spans and loads for wood structural panel sheathing (Based on IBC Table 2304.7(3))

Sheathing Grades		Maximum Span (inches)		Load (psf)	
Panel Span Rating	Panel Thickness	With Edge Support	Without Edge Support	Total Load	Live Load
16/0	3/8	16	16	40	30
20/0	3/8	20	20	40	30
24/0	3/8, 7/16, 1/2	24	20	40	30
24/16	7/16, 1/2	24	24	50	40
32/16	15/32, 1/2, 5/8	32	28	40	30

You Should Know

Engineered Wood Products

- Prefabricated I-Joists
- Structural Glued-Laminated Timber
- Structural Composite Lumber
  - Laminated Veneer Lumber (LVL)
  - Parallel Strand Lumber (PSL)

Code (IMC) requires the access opening to be large enough to remove the largest appliance (see Figure 16-23). **[Ref. 1209]**

Engineered wood products

Engineered wood products have gained popularity in construction in recent years. Several types of engineered wood products are available. Most of the products are manufactured by joining and/or bonding smaller-size wood materials to form a strong structural member. The most common engineered wood product is prefabricated wood joists. These joists are used extensively in both residential and commercial construction. The IBC permits the use of these materials in structures. They must be designed and installed in accordance with proven engineering practices and applicable referenced standards. They must be installed in accordance with the manufacturer's installation standards (see Figure 16-24).

Trusses

Trusses are floor and roof framing members comprised of light-frame materials (2 by 4 and 2 by 6) typically joined together with metal

FIGURE 16-23 Attic access must be large enough to remove mechanical equipment

FIGURE 16-24 I-joists and LVL beams frame second floor of apartment building

connector plates. Wood trusses must be designed in accordance with the IBC and accepted engineering practice. Engineered drawings of each truss used in a structure must be provided to the building official prior to installation as well as with the shipment of trusses to the job site. The trusses must be installed in accordance with the submitted truss drawings and the Truss Plate Institute's (TPI) National Design Standards for Metal-Plate-Connected Wood Truss Construction (see Figure 16-25).

FIGURE 16-25 Flat-plate wood truss floor system

NONSTRUCTURAL MATERIALS

The previous section of this chapter discussed building materials used for the structure of a building. This section discusses the materials that do not provide any type of structural support to the building. These materials are typically used as interior finishes or components within the structure itself for insulation, fire protection, and so on. Two types of materials are discussed: gypsum board and plastic materials.

Gypsum board

Gypsum board is a very common material used in buildings. It provides a smooth interior surface, is easy to work with, and sometimes provides passive fire protection for the building. It originally replaced the lath-and-plaster method of doing interior finishes. Gypsum board is manufactured by creating a panel of noncombustible gypsum core with paper facing on each side. It is used for wall and ceiling finishes in buildings. It is then taped and finished to produce a smooth flat surface. There are many different thicknesses and types of gypsum board. Type X and Type C gypsum boards have a higher level of fire resistance than regular gypsum board. Gypsum wallboard is not permitted to be used in any exterior surface where it will be exposed to the weather. It also cannot be used in areas where it will be directly exposed to water or conditions of high humidity (see Figure 16-26). **[Ref. 2508.2]**

FIGURE 16-26 Gypsum wallboard attached to steel studs

Gypsum board is applied in single or multiple levels. The most common is the single-ply application. The edges and ends of the gypsum board must be located on the framing members. It is then attached to the framing members by either nails or screws. The joints and fasteners in gypsum board are typically treated by applying a paper or fiberglass tape covered with joint-treatment compounds. However, the IBC only requires this treatment on fire-resistance-rated assemblies. The installation of the gypsum board must comply with the referenced standards and the manufacturer's installation instructions. **[Ref. 2508.3 and 2503.4]**

When gypsum board is used for a backing for tile, water-resistant gypsum backing board must be used. If wall tile is being installed in a shower or tub area, the backing must be glass mat water-resistant gypsum backing panels, discrete nonasbestos fiber-cement interior substrate sheets, or nonasbestos fiber-mat reinforced cement substrate sheets in compliance with ASTM Standard C 1178, Standard C 1288, or Standard C 1325. **[Ref. 2509.2]**

FIGURE 16-27 Foam plastic insulation covered with gypsum wallboard thermal barrier

Plastics

Although there are many types of plastics used in building and regulated by the IBC, the most common type is foam plastic insulation. Foam plastics used in buildings can produce rapid flame spread and heavy smoke. Therefore, the IBC regulates the installation of this product to limit these factors. Foam plastic insulation must have a flame-spread index of not more than 75 and a smoke-developed index of not more than 450 as tested in accordance with ASTM Standard E84 and Standard UL 723. **[Ref. 2603.3]**

When foam plastic is installed on the interior of a building, it must be separated from the interior by an approved thermal barrier. A thermal barrier is a single layer of ½-inch gypsum wallboard or other equivalent thermal barrier (see Figure 16-27). It must be installed so that it will remain in place for 15 minutes based on the referenced standards. There are several locations where foam plastic is permitted without the thermal barrier. Examples of those locations include cooler and freezer walls, walk-in coolers, roofs, doors with specific conditions, garage doors, and interior signs in covered mall buildings. **[Ref. 2603.4]**

Attics or crawl spaces where entry is made only for service of utilities may have foam plastic insulation without the thermal barrier. However, it must be protected against ignition by 1-inch-thick mineral fiber insulation; ¼-inch-thick wood structural panel, particleboard, or hardboard; ⅜-inch gypsum wallboard; corrosion-resistant steel having a base metal thickness of 0.016 inch; or other approved material installed in such a manner that the foam plastic insulation is not exposed. **[Ref. 2603.4.1.6]**

Foam plastic is often used as trim on the interior of a building. The IBC limits the amount of trim used to reduce the potential of increased flame spread and smoke development. The trim must have a minimum density of 20 pounds per cubic feet. It is limited to a maximum thickness of ½ inch and maximum width of 8 inches. The interior trim can constitute more than 10 percent of the wall or ceiling area to which it is attached. The flame-spread index for foam plastic trim is also limited to 75. **[Ref. 2604.2]**

GLOSSARY

A

addition – An extension or increase in floor area or height of a building or structure.

adhered masonry veneer – Veneer secured and supported through the adhesion of an approved bonding material applied to an approved backing.

alteration – Any construction or renovation to an existing structure other than repair or addition.

ambulatory health-care facility – Buildings or portions thereof used to provide medical, surgical, psychiatric, nursing, or similar care on a less-than-twenty-four-hour basis to individuals who are rendered incapable of self-preservation.

anchored masonry veneer – Veneer secured with approved mechanical fasteners to an approved backing.

approved – Acceptable to the code official or authority having jurisdiction.

area, building – The area included within surrounding exterior walls (or exterior walls and fire walls) exclusive of vent shafts and courts. Areas of the building not provided with surrounding walls shall be included in the building area if such areas are included within the horizontal projection of the roof or floor above.

B

building – Any structure used or intended for supporting or sheltering any use or occupancy.

building element – A fundamental component of building construction, which may or may not be of fire-resistance- rated construction and is constructed of materials based on the building type of construction.

building official – The officer or other designated authority charged with the administration and enforcement of the IBC, or a duly authorized representative.

built-up roof covering – Two or more layers of felt cemented together and surfaced with a cap sheet, mineral aggregate, smooth coating, or similar surfacing material.

C

closed system – The use of a solid or liquid hazardous material involving a closed vessel or system that remains closed during normal operations for which vapors emitted by the product are not liberated outside of the vessel or system and the product is not exposed to the atmosphere during normal operations. This includes all uses of compressed gases.

construction documents – Written, graphic, and pictorial documents prepared or assembled for describing the design, location, and physical characteristics of the elements of a project necessary for obtaining a building permit.

conventional light-frame construction – A type of construction for which the primary structural elements are formed by a system of repetitive wood-framing members.

court – An open, uncovered space, unobstructed to the sky, bounded on three or more sides by exterior building walls or other enclosing devices.

D

dwelling – A building that contains one or two dwelling units used, intended, or designed to be used, rented, leased, let, or hired out to be occupied for living purposes.

dwelling unit – A single unit providing complete, independent living facilities for one or more persons, including permanent provisions for living, sleeping, eating, cooking, and sanitation.

E

exit – That portion of a means of egress system that is separated from other interior spaces of a building or structure by fire-resistance-rated construction and opening protectives as required providing a protected path of egress travel between the exit access and the exit discharge.

exit access – That portion of a means of egress system that leads from any occupied portion of a building or structure to an exit.

exit discharge – That portion of a means of egress system between the termination of an exit and a public way.

F

fire area – The aggregate floor area enclosed and bounded by fire walls, fire barriers, exterior walls, or horizontal assemblies of a building. Areas of the building not provided with surrounding walls shall be included in the fire area if such areas are included within the horizontal projection of the roof or floor next above.

fire-resistance rating – The period of time a building element, component, or assembly maintains the ability to confine a fire, continues to perform a given structural function, or both, as determined by the tests or the methods based on tests.

fire separation distance – The distance measured from the building face to one of the following:

1. the closest interior lot line;
2. the centerline of a street, an alley, or public way; or
3. an imaginary line between two buildings on the property.

The distance shall be measured at right angles from the face of the wall.

G

guard – A building component or a system of building components located at or near the open sides of elevated walking surfaces that minimizes the possibility of a fall from the walking surface to a lower level.

H

habitable space – A space in a building for living, sleeping, eating, or cooking. Bathrooms, toilet rooms, closets, halls, storage or utility spaces, and similar areas are not considered habitable spaces.

handrail – A horizontal or sloping rail intended for grasping by the hand for guidance or support.

height, building – The vertical distance from grade plane to the average height of the highest roof surface.

high-rise building – A building with an occupied floor located more than seventy-five feet above the lowest level of fire department vehicle access.

J

jurisdiction – The governmental unit that has adopted the IBC under due legislative authority.

L

lot – A portion or parcel of land considered as a unit.

lot line – A line dividing one lot from another or from a street or any public place.

M

means of egress – A continuous and unobstructed path of vertical and horizontal egress travel from any occupied portion of a building or structure to a public way. A means of egress consists of three separate and distinct parts: the exit access, the exit, and the exit discharge.

O

occupant load – The number of persons for which the means of egress of a building or portion thereof is designed.

open system – Vessel or system that is continuously open to the atmosphere during normal operations and from which vapors are liberated or the product is exposed to the atmosphere during normal operations.

owner – Any person, agent, firm, or corporation having a legal or equitable interest in the property.

P

permit – An official document or certificate issued by the authority having jurisdiction that authorizes performance of a specified activity.

prefabricated I-joist – Structural member manufactured using sawn or structural composite lumber flanges and wood structural panel webs bonded together with exterior exposure adhesives, which forms an "I" cross-sectional shape.

primary structural frame – The primary structural frame shall include all of the following structural members:

1. the columns;
2. structural members having direct connections to the columns, including girders, beams, trusses, and spandrels;
3. members of the floor construction and roof construction having direct connections to the columns; and
4. bracing members that are essential to the vertical stability of the primary structural frame under gravity loading and that shall be considered part of the primary structural frame whether or not the bracing member carries gravity loads.

R

repair – The reconstruction or renewal of any part of an existing building for the purpose of its maintenance.

S

secondary members – The following structural members shall be considered secondary members and not part of the primary structural frame:

1. structural members not having direct connections to the columns;
2. members of the floor construction not having direct connections to the columns; and
3. bracing members other than those that are part of the primary structural frame.

shaft – An enclosed space extending through one or more stories of a building, connecting vertical openings in successive floors, or floors and roof.

sleeping unit – A room or space in which people sleep, which can also include permanent provisions for living, eating, and either sanitation or kitchen facilities but not both. Such rooms and spaces that are also part of a dwelling unit are not sleeping units.

stair – A change in elevation, consisting of one or more risers.

stairway – One or more flights of stairs, either exterior or interior, with the necessary landings and platforms connecting them, to form a continuous and uninterrupted passage from one level to another.

structure – That which is built or constructed.

T

through penetration – An opening that passes through an entire assembly.

through-penetration fire-stop system – An assemblage of specific materials or products that are designed, tested, and fire-resistance rated to resist for a prescribed period of time the spread of fire through penetrations.

V

ventilation – The natural or mechanical process of supplying conditioned or unconditioned air to or removing such air from any space.

W

water-resistive barrier – A material behind an exterior wall covering that is intended to resist liquid water that has penetrated behind the exterior covering from intruding farther into the exterior wall assembly.

wheelchair space – A space for a single wheelchair and its occupant.

wood structural panel – A panel manufactured from veneers, wood strands, or wafers or a combination of veneer and wood strands or wafers bonded together with waterproof synthetic resins or other suitable bonding systems.

Y

yard – An open space, other than a court, unobstructed from the ground to the sky, except where specifically provided by the IBC, on the lot on which a building is situated.

INDEX

Information in figures and tables is denoted by *f* and *t*.

A

B

G

H

I

O

P

R